INTERNATIONAL SERIES OF MONOGRAPHS ON
PURE AND APPLIED BIOLOGY

Division: **PLANT PHYSIOLOGY**

GENERAL EDITORS: P. F. WAREING and A. W. GALSTON

VOLUME 3

THE GERMINATION
OF SEEDS

OTHER TITLES IN THE PLANT
PHYSIOLOGY DIVISION

General Editors: P. F. WAREING and A. W. GALSTON

OTHER DIVISIONS IN THE SERIES ON
PURE AND APPLIED BIOLOGY

BIOCHEMISTRY

BOTANY

MODERN TRENDS
IN PHYSIOLOGICAL SCIENCES

PLANT PHYSIOLOGY

THE
GERMINATION
OF SEEDS

by
A.M. MAYER
and
A. POLJAKOFF-MAYBER

*Botany Department Hebrew University
Jerusalem*

A Pergamon Press Book

THE MACMILLAN COMPANY
NEW YORK
1963

THE MACMILLAN COMPANY
60 Fifth Avenue
New York 11, N.Y.

This book is distributed by
THE MACMILLAN COMPANY
pursuant to a special arrangement with
PERGAMON PRESS LIMITED
Oxford, England

Library of Congress Catalog Card Number 62-11557

Made in Great Britain

CONTENTS

PREFACE

ALTHOUGH this volume on germination is one of a series of monographs on subjects in plant physiology, we have not attempted to cover the entire field of germination. The number of papers is vast and goes back fifty to a hundred years. For this reason we have cited those papers which appeared to us as of importance. No doubt important papers have been omitted and differences of opinion are possible both as regards the selection and the arrangement of the material. An attempt has been made to treat the available information critically and to arrange it in an integrated form. At the same time it is impossible to avoid personal predilections and this has resulted in a more extensive treatment of those chapters dealing with subjects of special interest to the authors.

Our thanks are due to many scientists who have given us permission to use figures, tables and data from their published works, as follows: Professor P. Maheshwari and the McGraw-Hill Book Co. for Fig. 1.1, Professor W. Troll and the Gustav Fisher Verlag for Fig. 1.3 (a) and (b). Professor R. M. McLean and Longmans Green and Co. Ltd. for Fig. 1.3 (g), MacMillan and Co. Ltd. for Figs. 1.3 (b), (f) and (h), and Fig. 1.4 (a) from Hayward: *Structure of Economic Plants,* Dr. E. E. Conn and the *Journal of Biological Chemistry* for Table 5. 9, as well as to Drs. Y. Oota, M. Yamada, J. E. Varner, Y. Tazakawa, H. Albaum, S. P. Spragg, E. W. Yemm, H. Halvorson, A. Marcus, S. B. Hendricks, H. B. Sifton, S. Isikawa, P. F. Wareing, M. Holden, S. Brohult and A. M. McLeod for the material as cited in the text.

Our thanks are also due to others who have helped us in the preparation of the material.

Special thanks are due to Professor A. Fahn for reading and commenting on Chapter 1, and to the late Professor Hestrin and Dr. A. Lees for reading and commenting on Chapter 8, which is an attempt to demonstrate that the behaviour of seeds in their natural environment is not an isolated phenomenon in biology.

Lastly we are greatly indebted to our editor, Professor P. F. Wareing for his critical reading of the entire manuscript and for his many comments and suggestions, which helped to give the book its present form. Nevertheless, it goes without saying that the views expressed and the attitude taken are entirely our responsibility.

A. M. M.
A. P.-M.

Jerusalem

CHAPTER 1

THE STRUCTURE OF SEEDS AND SEEDLINGS

THE term germination is used to refer to a fairly large number of processes, including the germination of seeds, and of spores of bacteria, fungi and ferns as well as the processes occurring in the pollen grain when the pollen tube is produced. Although all these are processes of germination, we will confine the use of the term germination to the seeds of higher plants, the Angiosperms. Extension beyond this would lead to discussions which are well outside the scope of this monograph.

The seed of Angiosperms is essentially simple in structure and develops from a fertilized ovule. It consists of an embryo surrounded by an envelope. The embryo is usually derived from the fusions of nuclei of the male and female gametes, i.e. it is the result of the fertilization of the egg cell in the embryo sac by one of the male nuclei from the pollen tube. The envelope or testa originates from the mother plant and normally develops from the integuments of the ovule. Many seeds contain in addition an endosperm. The endosperm is derived from both parent plants in the case of cross-pollination, and from the same plant in self-pollination. It is formed as a result of the triple fusion of the two polar nuclei with the second sperm nucleus, Fig. 1.1. In some plants seeds arise by apomixis or non-sexual processes, the seed being formed usually from a diploid nucleus. Cases of polyembryony can arise either due to the division of the zygotic embryo, as in orchids, or due to secondary embryo development from the nucellar tissue, as in some citrus species. Various other aberrations in seed formation are also known.

1

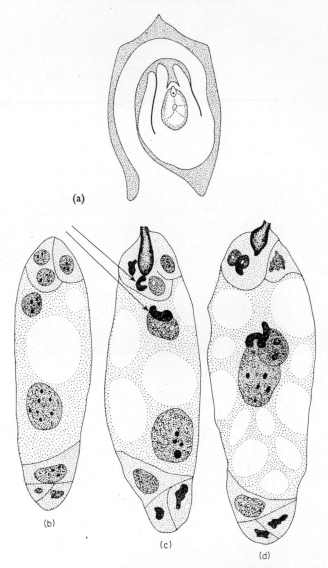

FIG. 1.1. Ovule and embryo sac before and after fertilization.

a.   General appearance of ovule of *Plumbago capensis*
b, c, d.   Embryo sac of *Lilium martagon* before and after fertilization
b.   Mature embryo sac
c.   Discharge of pollen tube into embryo sac (male nuclei)
d.   Contact of one male nucleus with egg nucleus and the other with two polar nuclei (after Maheshwari, 1950)

In addition to these three basic structures, embryo, testa and endosperm, other tissues may occasionally participate in the seed's make-up, which is by no means constant. The testa is sometimes made up of tissues of the parent plant other than the integuments, e.g. the nucellus, the endosperm and sometimes even the chalaza. The testa itself varies greatly in form. It may be soft, gelatinous or hairy, although a hard testa is the form most commonly met.

The size and shape of seeds is extremely variable. It depends on the form of the ovary, the condition under which the parent plant is growing during seed formation and, obviously, on the species. Other factors which determine the size and shape of seeds are the size of the embryo, the amount of endosperm present and to what extent other tissues participate in the seed structure. Some of the variety in seed form is illustrated in Fig. 1.2.

The normal seed contains materials which it utilizes during the process of germination. These are frequently present in the endosperm, as in *Ricinus*, tomato, maize and wheat (Fig. 1.3). The endosperm may contain a variety of storage material such as starch, oils, proteins or hemicelluloses. But an endosperm is by no means invariably present nor is it always the chief location of reserve materials. In many plants the endosperm is greatly reduced, as in the Cruciferae, and in the orchids its formation is entirely suppressed. In these cases, the reserve materials are present elsewhere, for example in the cotyledons of the embryo, as in beans and lettuce (Fig. 1.3). In the orchids this is not the case. They contain virtually no reserve materials and therefore represent a very special case of seed structure. In some plants the storage materials are contained in the perisperm, or the seed may contain both endosperm and perisperm, e.g. *Beta vulgaris* (Fig. 1.3). The perisperm originates from the nucellus, e.g. in the Caryophyllaceae and in *Coffea*, and not from the embryo sac as is the case for the endosperm.

Seeds are formed in the ovary and this develops into the fruit. Fruits arising primarily from the ovary are known as

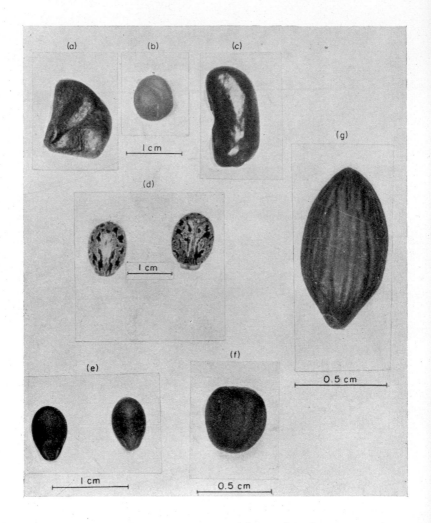

FIG. 1.2. Illustration of forms of seeds.

(a) *Pancratium maritimum*   (e) *Citrullus colocynthis*
(b) *Pisum sativum*   (f) *Raphanus sativus*
(c) *Phaeseolus vulgaris*   (g) *Xanthium pennsylvanicum*
(d) *Ricinus communis*

(a), (e) after Troll, 1954; (b), (f), (h) after Hayward, 1938; (g) after McLean
and Ivymey Cook, 1956).

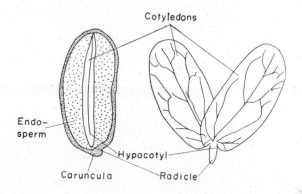

(a) Castor bean *(Ricinus communis)*

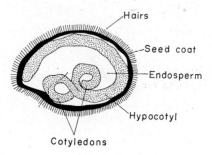

(b) Tomato

FIG. 1.3.(a, b) Structure of various seeds and of one seeded fruit, in which seed and fruit coats are fused.

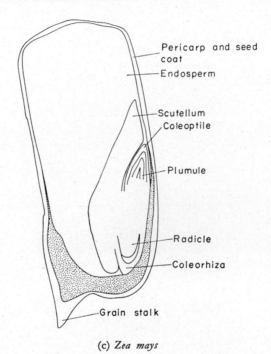

(c) *Zea mays*

Fig. 1. 3. (c) Structure of various seeds and of one seeded fruit, in which seed and fruit coats are fused.

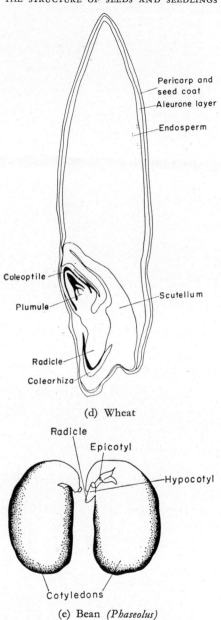

(d) Wheat

(e) Bean *(Phaseolus)*

Fig. 1. 3. (d, e) Structure of various seeds and of one seeded fruit, in which seed and fruit coats are fused.

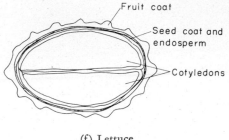

(f) Lettuce

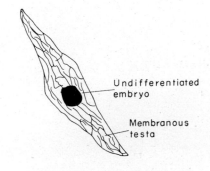

(g) *Odontoglossum* (Orchid)

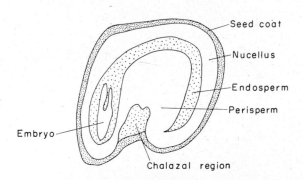

(h) Beet *(Beta vulgaris)*

FIG. 1.3. (f, g, h) Structure of various seeds and of one seeded fruit,
in which seed and fruit coats are fused.

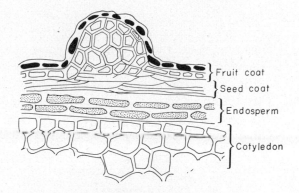

FIG. 1.4. Fusion of seed and fruit coat in lettuce and barley "seeds".

(a) Lettuce (after Hayward, 1938)

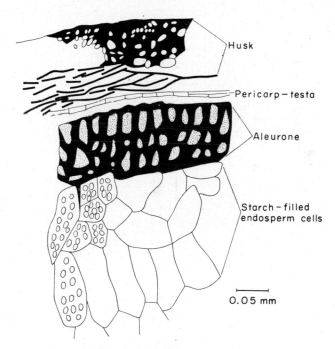

(b) Barley (after McLeod, 1960)

"true" fruits, while fruits in which other structures or several ovaries and their related structures participate, are often termed "false". Both types of fruits may be dry or fleshy. In many plants the integuments and the ovary wall are completely fused, so that the seed and fruit are in fact one entity as is the case in the grains of the grasses and in lettuce (Fig. 1.4). In other cases additional parts of the plants, such as glumes, bracts and neighbouring sterile florets remain attached to the fruit and form a bigger dispersal unit. The exact nature and classification of these organs need not concern us here. A few of the very variable seed-bearing structures are however illustrated in Fig. 1.5.

The embryo consists of a radicle, a plumule or epicotyl, one or more cotyledons and a hypocotyl which connects the radicle and the plumule. The embryo may be variously located within the seed and may either fill the seed almost completely, as in the Rosaceae and Cruciferae, or it may be almost rudimentary, as in the Ranunculaceae. The classification of seeds is often based on the size and position of the embryo and on the ratio of the size of the embryo to that of the storage tissue.

The process of germination leads eventually to the development of the embryo into a seedling. Seedlings are classified as *epigeal*, in which the cotyledons are above ground and are usually photosynthetic, and *hypogeal* in which the cotyledons remain below ground. In this latter case the cotyledons are either themselves the source of reserves for the seedling or receive materials from the endosperm. The size, shape and re-

---

FIG. 1.5. Illustration of various forms of dry fruits and dispersal units.

(a) *Medicago rotata*	(h) *Xanthium pennsylvanicum*
(b) *Lactuca sativa* var. Progress	(i) *Triticum vulgare*
(c) *Heliathus annuus*	(j) *Avena sativa*
(d) *Zea mays*	(k) *Lactuca sativa* var. Grand Rapids
(e) *Rumex rosea*	(l) *Papaver orientalis*
(f) *Erodium maximum*	(m) *Quercus ithaburensis*
(g) *Tipuana tipu*	(n) *Atriplex halimus*

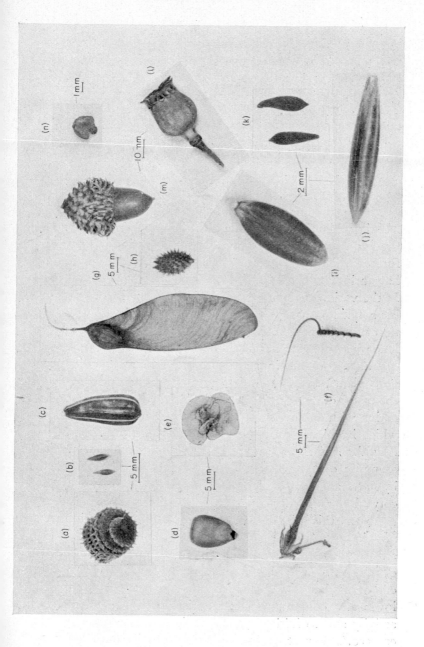

lative importance of the various tissues of seedlings are as variable as those of the seeds and fruits, and we will not attempt to classify them here. A few typical seedling forms are shown in Fig. 1.6. These indicate the large morphological variability which is met during germination.

In many plants special absorbing tissues develop which take up the reserve materials from the endosperm. In the onion the tip of the cotyledon is actually embedded in the endosperm and withdraws the materials from it (Fig. 1.6). The cotyledon with the seed attached is eventually raised above the ground and becomes green. In *Tradescantia* also the cotyledon is embedded in the endosperm, but the emerging organ is the coleoptile which is partly joined to the cotyledon and partly to the hypocotyl. In the Gramineae specialization has proceeded further and the scutellum fulfils the special function of a haustorium, which withdraws substances from the endosperm. In the various palms the haustorium, formed from the tip of the cotyledon, has become greatly enlarged. In a few cases the tip of the primary root rather than the scutellum of the cotyledon becomes the haustorium.

Such complications are absent in most of the dicotyledonous plants. Some straightforward examples are provided by the castor oil bean, *Ricinus communis*, which has epigeal germination with an endosperm, and *Phaseolus vulgaris* (French bean) which is epigeal without endospermic nutrition. In contrast, *Phaseolus multiflorus* is hypogeal and non-endospermic, while examples of hypogeal endospermic germination are provided by *Hevea* among the dicotyledons and *Zea mays* by the monocotyledons (Fig. 1.6).

After this brief discussion of forms of germination we are now in a position to try and define germination and to begin a discussion of the processes which lead up to it. Germination of the seed of the higher plant we may regard as that consecutive number of steps which causes a quiescent seed, with a low water content, to show a rise in its general metabolic activity and to initiate the formation of a seedling from the embryo. The exact stage at which germination ends and growth

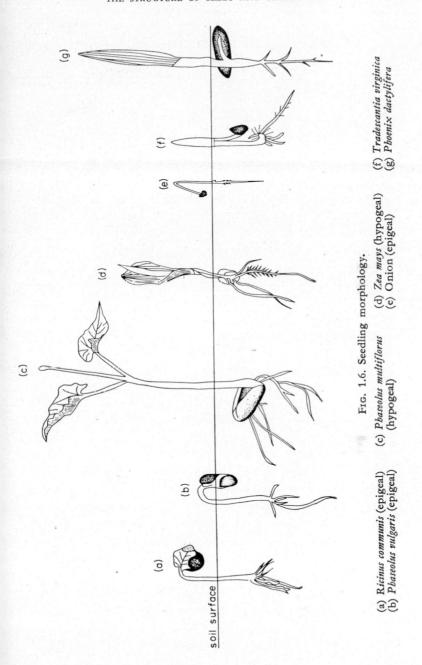

FIG. 1.6. Seedling morphology.

(a) *Ricinus communis* (epigeal)
(b) *Phaseolus vulgaris* (epigeal)

(c) *Phaseolus multiflorus* (hypogeal)

(d) *Zea mays* (hypogeal)
(e) Onion (epigeal)

(f) *Tradescantia virginica*
(g) *Phoenix dactylifera*

begins is extremely difficult to define. This is particularly difficult because we identify germination by the protrusion of some part of the embryo from the seed coat, which in itself is already a result of growth. There is no general rule as to which part of the embryo first pierces the seed coat. In many seeds this is the radicle and therefore germination is frequently equated with root protrusion. However in some seeds it is the shoot which protrudes first, for example in *Salsola*. The piercing of the seed coat by part of the embryo could be caused by cell division, cell elongation or both. In other words, the protrusion of part of the embryo through the seed coat is in fact the result of growth. Not all definitions would include cell division as such in the term growth. However, cell division is usually followed by an enlargement of the daughter cells and therefore constitutes growth. It is uncertain whether the process of germination can be equated with the processes of growth which lead to protrusion of part of the embryo through the seed coat. Probably the fundamental processes which cause germination are different from those of growth.

A further question which arises is whether cell division or cell elongation is the first process which occurs when the embryo pierces the seed coat. In some seeds it has been shown that cell division occurs first and is followed by cell elongation, while in other cases the reverse has been observed. Thus in lettuce seeds which are germinated at 26°C, cell division begins after some 12–14 hours and at about the same time cell elongation in root cells can also be observed (Evenari *et al.*, 1957).

In cherry seeds the changes in cell number and length of embryonic organs was followed during after-ripening under conditions of stratification at 5°C (Pollock and Olney, 1959). The length of the embryonic axis increased by some 7 per cent during stratification at 5°C. This increase in length was attributed to a combination of cell division and elongation. Changes at 5° and 25°C were similar for the first four weeks of stratification but following this, axis length increased much more at 5°C than at 25°C. At the latter temperature no germination was obtained. Thus cherry and lettuce seem to be similar with regard to the more or less simultaneous occurrence of elongation and division. In germinating seeds of *Pinus thunbergii*, on the other hand, it seems that cell division precedes cell elongation, but the evidence is not entirely clear (Goo, 1952). Cell division occurs during a period when there is very little water uptake and before germination.

The reverse observation has been made in germination of *Zea mays*, where the first change is cell enlargement in the coleorhiza. Cell division in the radicle occurs later, when the latter breaks through the coleorhiza and the seed coat. It is the coleorhiza which first pierces the seed coat (Toole, 1924). A similar situation apparently exists in barley.

A further attempt to differentiate between division and elongation during germination of lettuce has been made by Haber and Luippold (1960). Gamma radiation and low temperatures were used to delay cell division and hypertonic solutions of mannitol to arrest cell elongation. Despite the fact that irradiation and low temperatures prevented mitosis, germination as observed by root protrusion occurred. When, however, cell elongation was prevented by hypertonic solutions, germination was prevented, although cell division could be clearly noted. Haber and Luippold conclude that division and cell elongation are affected by different factors and that root protrusion is primarily the result of cell elongation. Cell division only serves to increase the number of cells which can elongate. It is by no means clear whether these observations also apply to other seeds.

Cell division and cell elongation only occur, once the cells of the seed have been activated in some way, so as to permit the control of these processes by various factors. Although it appears that for visible germination only one of these processes is necessary, there can be no doubt that for normal development and growth of the seedling both cell division and cell elongation are essential. It is thus possible to differentiate between these growth processes and the process of activation, which precedes growth and which may be termed germination.

However we will not, during the discussion of germination, attempt to make such a precise differentiation but rather will try to describe all those processes which take place up to seedling formation.

BIBLIOGRAPHY

EVENARI, M., KLEIN, S., ANCHORI, H. and FEINBRUNN, N. (1957) *Bull. Res. Council, Israel,* **6** *D*, 33.

GOO, M. (1952) *J. of. Jap. Forestry Soc.* **34**, 3.

HABER, A. H. and LUIPPOLD, H. J. (1960) *Plant Phys.* **35**, 168.

HAYWARD, H. E. (1938) *The Structure of Economic Plants,* Macmillan, New York.

MCLEAN, R. M. and IVIMEY-COOK, W. R. (1956) *Textbook of Theoretical Botany,* vol. 2, Longmans Green, London.

MCLEOD, A. M. (1960) *Wallerstein. Lab. Comm.* **23**, 87.

MAHESHWARI, P. (1950) *An Introduction to the Embryology of Angiosperms,* McGraw-Hill, New York.

POLLOCK, B. M. and OLNEY, H. O. (1959) *Plant Phys.* **34**, 131.

TOOLE, E. H. (1924) *Amer. J. Bot.* **11**, 325.

TROLL, W. (1954) *Praktische Einführung in die Pflanzenmorphologie,* Gustav Fisher, Jena.

CHAPTER 2

CHEMICAL COMPOSITION OF SEEDS

THE chemical composition of seeds shows the same variability as is seen in other plant characteristics. The compounds found can be divided roughly into two groups, viz. (1) the normal constituents which are likely to occur in every plant tissue and (2) storage materials, which are frequently present in seeds in very large amounts. In addition, a large number of secondary plant products may be present in seeds. Most of the compounds which occur in them are not normally different from those found in the plant, but seed proteins differ in chemical composition and properties from those found in other plant tissues. The occurrence of large quantities of lipids differentiates seeds from all other plant tissues except certain fruits, since lipids do not usually occur in large amounts in other plant tissues.

Seeds can be divided into those whose main storage material is carbohydrate and those whose main storage material is lipid. Lipid-containing seeds are by far the bigger of these two groups, although among economically important seeds this preponderance does not occur to the same extent. Seeds containing proteins can belong to either group. Almost no seeds are known in which the predominant storage material is protein, although soybeans are an exception. Precise information about seed composition is available chiefly for seeds which are used either for food or in industry. The chemical composition of seeds of various vegetable crops is less well known, while the information about the composition of the seeds of wild plants is extremely scant.

The constituents of seeds are determined genetically, but the relative amounts of these constituents are sometimes dependent on environmental factors such as mineral nutrition and climate. Thus Iwanoff (1927) showed that the protein content of wheat varied in various parts of Russia while that of peas was constant (Table 2. 1). Modern plant breeding practice has permitted selection for quantitative differences in seed constituents. Thus soybeans have been bred for high protein content, flax for oil content and wheat both for protein and starch content. In some interesting breeding experiments, Woodworth *et al.* (1952) selected maize for various contents of protein and oil. Starting with maize containing 4·7 per cent oil and 10·9 per cent protein they obtained, after fifty generations of selection, four varieties having 15·4 or 1·0 per cent oil and 19·5 or 4·9 per cent protein. Tables 2.1, 2.3 and 2.4 illustrate some of the variability in seed composition. An example of the composition of a wild plant, *Amaranthus*, is shown in Table 2.5. From these tables it will be seen that in addition to carbohydrates, proteins and lipids seeds contain minerals, tannins, phosphorous compounds, etc. A classification of seed constituents is difficult because of the large variety of

TABLE 2.1 — PROTEIN CONTENT OF WHEAT *(Triticum vulgare*, var. *albidum)* AND PEAS *(Pisum sativum*, var. *vulgare)* GROWN IN VARIOUS PARTS OF RUSSIA AND HARVESTED IN 1924
(After Iwanoff, 1927)

Place	Longitude	Latitude	Protein content % dry weight	
			Wheat	Peas
Severo Dvinsk	46°18'	60°46'	11·92	27·00
Moscow	37°20'	55°48'	14·30	26·56
Kiev	30°28'	50°27'	19·32	30·37
Saratov	45°45'	51°37'	21·01	30·37
Omsk	70°32'	55°11'	18·69	28·12
Krasnojarsk	92°52'	56°01'	19·03	26·06
Vladivostok	131°57'	43°05'	11·86	26·87

TABLE 2.2 — CHEMICAL COMPOSITION OF SEEDS
(After Wehmer, 1929, Anderson and Kulp, 1921; Gutlin Schmitz, 1957;
and Czapek, 1905)

	Percentage of air-dry seeds				
	Carbohydrates		Proteins	Fats	Lecithin
	Starch	Sugar			
Zea mays	50–70	1–4	10.0	5	
Pisum sativum	30–40	4–6	20.0	2	1.2
Arachis hypogea	8–21	4–12	20–30.0	40–50	
Helianthus annuus	0	2	25.0	45–50	
Ricinus communis	0	0	18.0	64	
Acer saccharinum	42	20	27.5	4	
Triticum	60–75		13.3	2.0	0.7
Fagopyrum esculentum		72.0	10.0	2.0	0.5
Chenopodium quinoa		48.0	19.0	5.0	
Aesculus hippocastanum		68.0	7.0	5.0	
Castanea vesca		42.0	4.0	3.0	
Quercus pendunculata		47.0	3.0	3.0	
Linum usitatissimum		23.0	23.0	34.0	
Brassica rapa		25.0	20.0	34.0	
Papaver somniferum		19.0	20.0	41.0	0.4
Cannabis sativa		21.0	18.0	33.0	0.9
Amygdalus communis		8.0	24.0	53.0	
Aleurites moluccana		5.0	21.0	62	

TABLE 2.3 — CHEMICAL COMPOSITION OF
LETTUCE SEEDS (Original)

	air-dry seeds (mg/g)
Total dry weight	960.0
Ash	46.0
Phytic Acid	20.0
Sucrose	30.0
Glucose	2.0
Fat	370.0
Total nitrogen	40.0
Protein nitrogen	37.0
Soluble nitrogen	1.0
Riboflavin	0.012
Ascorbic acid	0.029
Carotene	0.004
Total P (free and bound)	8.5–14

TABLE 2.4 — CHEMICAL COMPOSITION OF SOYBEANS
(Compiled from data of Morse, 1950)

A. Major Constituents   % on moisture-free basis

Moisture	8·0
Ash	4·6
Fat	18·0
Fibre	3·5
Protein	40·0
Pentosans	4·4
Sugars	7·0
Starch-like substances	5·6
Phosphorus	0·63
Potassium	1·67
Calcium	0·26

B. Mineral Constituents   % air-dry weight of seeds

Magnesium	0·22
Sulphur	0·41
Chlorine	0·024
Iodine	trace
Sodium	0·34
Manganese	0·0028
Zinc	0·0022
Aluminium	0·0007
Copper	0·0012
Iron	0·0097

C. Vitamins   µg/g

Thiamine	17·5
Riboflavine	3·6
Pyridoxine	11·8
Nicotinic Acid	21·4
Pantothenic Acid	21·5
Inositol	2291·0
Biotin	0·8

TABLE 2.5 — COMPOSITION OF SEEDS OF
Amaranthus retroflexus
(Woo, 1919)

	% air-dry seeds
Water	8·6
Lipids	7·8
Polysaccharides	47·2
Reducing sugars	none
Non-reducing sugars (after hydrolysis)	1·2
Nitrogen	2·5
Protein	15·0 (Soluble 3·0 Insoluble 12·0)
Ash	4·2

compounds met with, but a rough breakdown into carbohydrates, proteins, lipids and other compounds seems permissible.

I. CARBOHYDRATES

The two chief storage carbohydrates are starch and hemicelluloses. The former is found in all the food grains and in legumes, while hemicelluloses, both pentosan and hexosan, occur in the endosperm of the palms as well as in the cotyledons of Lupins, *Primula* and *Impatiens*. In addition, however, many other carbohydrates occur in seeds, not necessarily as storage materials. Thus in many seeds various polyuronides occur frequently in the seed coat. These mucilages are possibly connected with seed dispersal and water uptake during germination.

Mucilages occur either on the seed surface, as in flax, or in special cells in the seed coat, as in *Brassica alba*. *Plantago* seeds constitute a commercial source for mucilages. Chemically the mucilages are polyuronides, mainly galacturonides. In addition, various sugars, hexoses and pentoses, have been isolated from mucilages. The polyuronides are frequently associated with proteins. Occasionally cellulose fibres are found in muci-lages.

Pectins are a normal constituent of plant cells and conse-quently also of seeds. Mono-, di-, tri- and oligo-saccharides, e.g. glucose, fructose, sucrose, raffinose and stachyose, occur in greater or smaller amounts in most seeds. In barley a consider-able amount of fructosans occurs. The carbohydrate com-position of barley seeds is shown in Table 2.6, as is the divi-

TABLE 2.6 — CARBOHYDRATE COMPOSITION OF BARLEY
(McLeod, 1960)

% dry-weight of tissue

	Husk	Embryo	Endosperm
Sucrose	0	14·0	0·2
Raffinose	0	10·0	0·08
Hexoses	0	0·2	0·26
Total glucan	<0·02	0	1·7
Total pentosan	3·6	0·4	1·0
Galactan	0	0·3	0
Uronic acid	+	+	0
Crude cellulose	30	7·0	0·4

sion of these constituents among various parts of the seed. Ivory nuts contain mannans, and galactans occur in *Lupinus albus* and *Strychnos*. Many secondary plant products are present in seeds in the form of glycosides. Thus bitter almonds con-tain amygdalin, mandelo-nitrile gentiobioside. Black mustard seeds contain sinigrin, the glycoside of a mustard oil, and *Nigella* contains damasenine. Various alkaloids, tannins and leucoanthocyanins also occur as glycosides in seeds.

II. Lipids

Lipids are generally present in the form of the glycerides of fatty acids: $CH_2 \cdot OR_1$

$$CHOR_2$$

$$CH_2 \cdot OR_3$$

where R_1, R_2 and R_3 may be the same, or different fatty acids. Most seed fatty acids are unsaturated and the most commonly-occurring ones are oleic, linoleic and linolenic acids. In addi-

TABLE 2.7 — COMPOSITION OF LIPIDS FROM SOYBEANS
(Compiled from data of Daubert, 1950)

	Total	% dry weight of bean
Saturated Fatty Acids	11–13	
Myristic		0·1–0·4
Palmitic		6·5–9·8
Stearic		2·4–5·5
Arachidic		0·2–0·9
Unsaturated Fatty Acids	88	
Oleic		11–60
Linoleic		25–63
Tetradecenoic		trace
Hexadecenoic		trace
Average Iodine Value	102–150	

Other Constituents
  Carotene (chiefly β)
  Chlorophyll
  Sterols
  Phytosterolins
  Tocopherols
  Phosphatides

tion, however, many other organic acids, both saturated and unsaturated, such as acetic, butyric, palmitic, stearic, lauric and myristic acids and many others, occur as glycerides. Ground-nuts contain the glyceride of arachidic acid. The lipids are found both as fats and as oils, depending on the relative amounts of saturated and unsaturated fatty acids occurring in the glycerides. Other lipid materials found in seeds are esters of higher alcohols, and sterols and phospholipids of various kinds. An example of the composition of seed lipids from soybeans is given in Table 2.7.

A highly toxic oil has been isolated from seeds of *Dichapetalum toxicarium*. This lipid has been shown to contain fluoro-oleic acid (Peters *et al.*, 1960). The compound is so toxic that the seeds of this plant are often referred to as "ratsbane". It is interesting that the leaves of this plant contain fluoro-acetic acid.

III. PROTEINS

One of the characteristics of proteins of seeds is that while some of them are metabolically active, such as the enzyme proteins and nucleoproteins, a large part of them are metabolically inactive. These latter are the storage proteins which vary according to the species. Thus, in wheat, at least four different proteins occur, glutelins, prolamins, globulins and albumins. The glutelins and prolamins form the major component of the protein, while the globulins contribute only 6–10 per cent of the total and the albumins 3–5 per cent. The total active proteins, globulins and albumins account for no more than 15 per cent of the total proteins in wheat. The distribution between metabolically-active and inactive proteins is similar in most cereals.

The division of the storage proteins into glutelins and prolamins is somewhat arbitrary and is based primarily on the differential solubility of the proteins in weak acids and alkalis. It is therefore preferable to refer to definite proteins whose properties have been investigated. The best-known prolamins are those found in wheat-gliadin, in barley-hordein and in maize-zein. Glutelin from wheat is probably a mixture of a number of proteins. Other glutelin proteins, such as zecanin, hordenine and oryzenin have not been studied by modern methods and it is therefore not certain whether single proteins or mixtures of proteins are involved. In dicotyledonous plants prolamins seem to be almost absent. Glutelins are sometimes absent and sometimes constitute up to 50 per cent of the total proteins in dicotyledonous seeds. Albumins and globulins in these seeds are usually well-defined and a number of the globulins have been obtained in crystalline form after extraction with hot salt solutions. Among those which have been investigated in detail are legumin and vicillin from peas, which have been obtained electrophoretically homogeneous, arachin and conarachin from peanuts, glycinin from soybeans and edestin from hemp seeds.

Seed storage-proteins generally have a high nitrogen content, high proline content and are often low in their content of ysine, tryptophan and methionine. Table 2.8 shows the pro-

TABLE 2.8 — THE PROTEIN CONTENT OF VARIOUS SEEDS
(From Bröhult and Sandegren, 1954)

	Total protein	Various fractions as % of total protein			
	% dry seeds	Albumin	Globulin	Prolamin	Glutelin
Triticum vulgare	10–15	3–5	6–10	40–50	30–40
Hordeum vulgare	10–16	3–4	10–20	35–45	35–45
Avena sativa	8–14	1	~80	10–15	~5
Secale cereale	9–14	5–10	5–10	30–50	30–50
Cucurbita pepo	12	very little	92	very little	small amounts
Nicotiana sp.	33	24	26	very little	50
Gossypium herbaceum	~20	very little	90	very little	10
Glycine hispida	30–50	small amounts	85–45	very little	very little
Lupinus luteus	~40	1	78	very little	16

tein content of some mono- and dicotyledonous seeds, and
Table 2.9 shows the amino acid composition of some seed
proteins.

TABLE 2.9 — AMINO ACID COMPOSITION OF SOME SEED PROTEINS
(Amino acid as % of protein)
(From Brohult and Sandegren, 1954; and Circle, 1950)

	Gliadin	Zein	Legumin	Glycinin	Arachin
Total nitrogen	17·0	16·2	17·9	17·0	18·3
Glycine	1·0	0		1·0	3·7
Alanine	2·0	11·5			5·0
Serine	4·8	7·8			5·3
Threonine	2·1	3·0			2·6
Valine	2·6	3·0		1·0	4·5
Leucine	6·7	24·0		8·5	7·8
Isoleucine	(5·1)	7·4		2·5	7·6
Methionine	1·6	2·3		1·8	0·7
Cysteine and/or cystine	2·5	1·0		1·1	1·5
Proline	13·2	10·5		4·0	7·0
Phenylalanine	6·3	6·5		4·0	6·3
Tryptophan	0	0	1·3	1·6	1·0
Tyrosine	3·3	5·3	4·3	1·8	5·5
Histidine	2·3	1·7	3·0	1·4	2·2
Arginine	2·7	1·8	13·1	8·1	14·0
Lysine	1·2	0	3·5	4·0	4·6
Aspartic acid	3·6	5·7	16·3	5·0	(16·0)
Glutamic acid	4·7	27·0	30·0	19·0	23·8

IV. OTHER COMPONENTS

In addition to the major compounds mentioned, seeds
contain a large number of other substances. All seeds contain a certain amount of minerals (see Table 2.4). The mineral
composition of seeds is essentially similar to that of the plant
and usually comprises all the essential and minor elements.
The phosphorous composition of *Amaranthus* is shown in
Table 2.10. A special feature of many seeds is that a very large

TABLE 2.10 — PHOSPHORUS COMPOUNDS IN
Amaranthus retroflexus
(Woo, 1919)

	P as % of total ash
Total P	4·6
Inorganic P	0·13
Lipid P	0·18
Soluble organic P	0·33
Phosphoprotein P	1·8
Nucleoprotein P	2·5

part of the phosphate occurs as phytin, the calcium and magnesium salt of inositol hexaphosphate. In lettuce seeds 50 per
cent of the total phosphorus occurs as phytin, 6–10 per cent as
free phosphate and the remainder in other phosphorus-containing compounds such as nucleotides, sugar-phosphates
(20–25 per cent), phospholipids, nucleoproteins and other compounds (20–25 per cent). In addition, the nucleic acids constitute an extremely important part of the phosphorus-containing compounds. The nucleic acids occur partly in their free
form and partly in the form of nucleoproteins. The ratio
of RNA to DNA (ribonucleic acid to deoxyribonucleic acid)
in many seeds is approximately 10 : 1.

The nitrogen content of seeds usually comprises, in addition to proteins, a certain amount of free amino acids and amides. The amides found are glutamine and asparagine, as well
as γ-methylene glutamine, as for example in ground-nuts. The
free amino acids found in seeds are usually the same as those
forming part of the protein structure. In addition, a few other
amino acids such as γ-methylene glutamic, γ-aminobutyric,
β-pyrazol-1-ylalanine and other heterocyclic amino acids have
also been found in certain seeds (Fowden, 1960; and Noe and
Fowden, 1960). It is not yet clear to what extent some of the
more recently discovered amino acids which occur in many
plants, also occur in the corresponding seeds.

Further nitrogenous constituents of seeds are various alkaloids. Some examples are provided by piperine in *Piper nigrum* seeds, ricinine in Castor oil beans, hyoscine in seeds of *Datura* and lupinidine (sparteine) in lupin seeds. It is interesting to note that the occurrence of some alkaloids in the plant does not necessarily indicate its presence in the seed in comparable amounts. A special case is that of *Coffea*, where the caffeine content of the bean is much higher than that of the plant. Cacao *(Theobroma cacao)* has a very high content of theobromine in the seeds as well as a smaller amount of caffeine. *Cola nitida* seeds also provide a source of caffeine. Strychnine and brucine are obtained from the seeds of *Strychnos nux-vomica* where they amount to about 2–3 per cent of the seeds.

Various organic acids such as tricarboxylic acid cycle intermediates, as well as malonic acid, have been detected in seeds of many species.

Phytosterols occur in a number of seeds. The best known ones are the sitosterols and stigmasterols from soybeans. The latter is important pharmaceutically as it is used as a precursor of progesterone.

Although seeds contain a number of pigments, chlorophyll is usually absent although it occurs in the seeds of gymnosperms. Protochlorophyll, however, occurs in the Cucurbitaceae. Breakdown products of chlorophyll seem to be present in a number of seeds. Other pigments found are carotene and various other carotenoids. The seed coats of many seeds contain anthocyanins or leucoanthocyanins. Flavonoid pigments are also known to be present in various seeds. Many of these compounds occur as the corresponding glycosides. Thus the seed coat of *Phaseolus vulgaris* may contain leuco-delphinidin, leuco-pelargonidin as well as delphinidin, petunidin and malvidin as the corresponding glycosides, as well as at least two flavonol glycosides. Cotton seeds contain a yellow pigment, gossypol, in special pigment glands.

Various phenolic compounds such as coumarin derivatives, chlorogenic acid and simple phenols such as ferulic, caffeic and sinapic acids, occur in many seeds. These compounds may give rise by oxidation and condensation to pigments of the melanin type. Another phenolic type constituent is the tannins, e.g. in *Arachis* seeds.

Although the function of vitamins in seeds is as yet unknown, in all cases where seeds have been examined vitamins have in fact been found to be present. The presence of vitamins is especially important in those seeds which are used for food or fodder. In these, detailed information on the vitamin content and especially the B component is available, and attempts are continuously made to raise the vitamin content of such seeds by selective breeding. Some figures are given in Table 2.11. Tocopherols are present in the oil of many seeds. Most of the known vitamins have in fact been shown to occur in some kinds of seeds.

With the development of modern analytical methods and paper chromatographic techniques, linked with biological assays, it has become possible to study the growth substance content of various seeds. The occurrence of indolylacetic acid in a number of seeds has been demonstrated, while in other

TABLE 2.11 — THE VITAMIN CONTENT OF SOME SEEDS. THE FIGURES ARE FOR THE AIR-DRY SEEDS
(From *Food Composition Tables*, FAO, 1954)

	mg/100 g				I. U.
	Thiamin	Ribo-flavin	Nicotinic Acid	Ascorbic Acid	Vit. A
Wheat (Durrum)	0·45	0·13	5·4	0	0
Rice (hulls removed)	0·33	0·05	4·6	0	0
Barley	0·46	0·12	5·5	0	0
Maize	0·45	0·11	2·0	0	450
Ground-nuts(shelled)	0·84	0.12	16·0	0	30
Soybean	1·03	0·30	2·1	0	140
Broad bean	0·54	0·29	2·3	4	100
Peas	0.72	0·15	2·4	4	100
Sunflower	0·12	0.10	1·4	0	30
Chestnuts (fresh)	0·21	0·17	0·4	24	0

cases various indole derivatives have been found. Gibberellic acid and gibberellin-like substances have been found in runner beans, in lettuce seeds and in the seeds of many other plants. In addition, growth-promoting and growth-inhibiting substances have been shown to be present in various seed extracts, but their nature has not yet been elucidated. The possible importance of these substances in the regulation of dormancy will be discussed in a later chapter.

BIBLIOGRAPHY

ANDERSON, R. J. and KULP, W. L. (1921) *N. Y. Agr. Expr. Sta. Bull.* **81**.

BROHULT, S. and SANDEGREN, E. (1954) In *The Proteins* vol. 2A, 487, Academic Press, New York.

CIRCLE, S. J. (1950) In *Soybeans and Soybean Products,* vol. 1, 275, Interscience, New York.

CZAPEK, F. (1905) *Die Biochemie der Pflanzen,* Gustav Fisher, Jena.

DAUBERT B. F. (1950) In *Soybeans and Soybean Products,* vol. 1, 157, Interscience New York.

FAO Food Composition Tables, Minerals and Vitamins, Rome, Italy (1954).

FOWDEN, L. (1960) *Rep. Progr. Chem.* **56**, 359.

GUTLIN SCHMITZ, P. H. (1957) Dissertation of the University of Basel: Über die Änderung des Aneuringehaltes während der Keimung in Samen verschiedenener Reserve Stoffe.

IWANOFF, N. N. (1927) *Biochem. Z.* **182**, 88.

IWANOFF, N. N. (1927) *Bull. Bot. and Plant Breeding* **17**, 225.

McLEOD, A. M. (1960) *Wallerstein Lab. Comm.* **23**, 87.

MORSE, W. J. (1950) In *Soybeans and Soybean Products,* vol. 1, 135, Interscience, New York.

NOE, F. F. and FOWDEN, L. (1960) *Biochem. J.* **77**, 543.

PETERS, R. A., HALL, R. J,, WARD, P. F. and SHEPPARD, N. (1960) *Biochem. J.* **77**, 17.

WEHMER, C. (1929) *Die Pflauzenstoffe* 2nd. Edition, Jena

WOO, M. L. (1919) *Bot. Gaz.* **68**, 313.

WOODWORTH, C. M., LENG, E. R. and JUGENHEIMER, R. W. (1952) *Agron. J.* **44**, 60.

CHAPTER 3

FACTORS AFFECTING GERMINATION

I. Viability and Life Span of Seeds

SEEDS are fairly resistant to extreme external conditions, provided they are in a state of desiccation. As a result seeds can retain their ability to germinate, or viability, for considerable periods. The length of time for which seeds can remain viable is extremely variable and depends both on the storage conditions and on the type of seed. In general, viability is retained best under conditions in which the metabolic activity of seeds is greatly reduced, i.e. low temperature and high carbon dioxide concentration. In addition, however, other factors are of great importance, particularly those which determine seed dormancy. The period for which seeds remain viable is determined genetically and by environmental factors. The latter will in fact have a decisive effect on the life span of any given seed, i.e. whether the seed will remain viable for the longest genetically possible period or whether it will lose its viability at some earlier stage.

Becquerel (1932 and 1934) tried to germinate seeds which were taken from the Herbaria of the National Museum in Paris, in order to estimate their life span. He came to the conclusion that *Mimosa glomerata* seeds remained viable for some 221 years, while various other leguminous seeds remained viable for periods of 100–150 years, e.g. *Astragalus massiliensis*, *Dioclea paucifera* and *Cassia bicapsularis*.

Turner (1933) tested seeds from collections kept at Kew. He also cites many examples in which life span was estimated from data on length of burial period and similar circumstantial

31

TABLE 3.1 — LIFE SPAN OF SEEDS IN YEARS
(From data of Becquerel 1932 and 1934; and Turner, 1933)

Anagallis foemina	60	Lotus uliginosus	81
Anthyllis vulneraria	90	Medicago orbicularis	78
Cassia bicapsularis	87	Nelumbium luteum	56
Cytisus biflorus	84	Stachys nepetifolia	77
Ipomoea sp.	43	Trifolium arvense	68
Lavateria olbia	64	Trifolium pratense	81
Lens esculenta	65	Trifolium striatum	90

information. Some of the data of Turner and Becquerel are given in Table 3.1. Many seeds (e.g. charlock) kept better, buried in soil, than when in jars on the laboratory shelves.

In contrast to the seeds listed in Table 3.1, which have a very long life span, other seeds are characterized by a very short life span. For example *Acer sacharinum, Zizana aquatica, Salix japonica* and *S. pierotti* lose their viability within a week if kept in air. *Ulmus campestris* and *U. americana* remain viable for about 6 months. *Hevea*, sugar cane and *Boehea, Thea*, cocos and other tropical crop seeds, remain viable for less than a year (Crocker, 1938).

Several attempts have been made to establish the viability of seeds by controlled experiments. The seeds were usually buried in the soil in some suitable containers and samples removed at different periods. Thus Beal began experiments in 1879 and these were continued for a period of 30 years, seeds being removed and tested every five years. After thirty years a considerable number of species was still viable, namely *Amaranthus retroflexus, Brassica nigra, Capsella bursa-pastoris, Lepidium virginicum, Oenothera biennis, Rumex crispus* and *Setaria media*. Similar experiments were conducted by Duvel (1905) and summarized by Goss in 1924. The final results of this experiment were reported by Toole and Brown (1946). A long term experiment planned to last more than three hundred years has been initiated by Went and Munz (1949).

Both in the experiments of Beal and those of Becquerel the seeds were maintained under relatively dry conditions during storage. It is known today that seeds generally remain viable for longer periods if they are dry. For example lettuce seeds kept much longer if their moisture content was reduced from air-dry to half this value in Ithaca (U.S.A.) (Griffiths, 1942). Moisture content was more critical than temperature during storage. Raising moisture content from 5 to 10 per cent caused a more rapid loss of viability than a temperature rise from 20° to 40°C. Similar data have been obtained for clover. Here the seeds remained viable for three years at all temperatures up to 38°C when their moisture content was 6 per cent. At 8 per cent moisture they remained viable except at this highest temperature. When the moisture content of the seeds was raised to 12 or 16 per cent the seeds remained viable only at 30°C. At 16 per cent moisture content viability at 22°C fell within three months (Ching *et al.*, 1959).

An interesting attempt has recently been made to express more exactly the relationship between viability, storage temperature and moisture content of cereal seeds (Roberts, 1960). Roberts was able to express all the known data on viability by a simple mathematical relationship:

$$\log p = K_v - C_1 m - C_2 t$$

where p is the half life viability period of the seeds, t is the temperature in °C, m the moisture as per cent and K and C are constants. These constants were calculated from experiments with wheat, but they appeared to fit also data obtained from oats and barley. From this expression it is possible to predict the expected viability and life span of a given cereal seed under almost all storage conditions.

Although from the above data it appears that dry conditions are essential for retention of viability, many seeds remain viable when submerged in water. Shull (1914) was able to show that 11 out of 58 species tested were still viable after $4\frac{1}{2}$ years submergence. Some of the results obtained by Shull are summarized in Table 3.2 which lists the species of seeds submerged under water and those subsequently found to germinate.

TABLE 3.2 — VIABILITY OF VARIOUS SEEDS WHEN KEPT SUBMERGED UNDER WATER FOR VARIOUS PERIODS OF TIME
(+ seeds germinated after indicated period of submergence) (Shull, 1914)

	130 days	18 months	30 months	54 months	7 years
Agrimonia hirsuta					
Asclepias syriaca				+	
Chenopodium album			+		
Circaea lutetiana					
Geum carolinianum					
Hieracium sp.					
Juncus bufonius	+				+
Juncus tenuis	+	+		+	+
Lappa minor			+		
Muhlenbergia diffusa		+	+	+	
Phryma leptostachya					
Plantago rugelii	+	+	+	+	
Polygonum arifolium					
Polygonum virginianum					
Rhus glabra					
Sanicula marylandica					
Sium cicutaefolium		+	+	+	+
Solidago rugosa				+	
Sparganium androcladum				+	
Unifolium canadense					
Verbena urticaefolia		+	+	+	
Washingtonia longistylis					

From the foregoing it can be seen that the storage conditions required to maintain viability for different seeds are different. Thus cases are known where drying causes very rapid loss in viability, e.g. *Acer saccharinum*, while in other cases only on drying will the seeds remain viable. Similar differences in storage conditions are known for oxygen and carbon dioxide concentration. Therefore it is very difficult to formulate any general rule for favourable storage conditions. This problem is also discussed by Owen (1956).

Even under favourable storage conditions many seeds are relatively short-lived, for example the seeds of many trees and of various vegetables. Such a loss of viability is not a sudden abrupt failure to germinate in all the seeds in a certain population. Rather the percentage of seeds which will germinate in any given population will slowly decrease. Moreover even if a seed loses its viability this does not imply that all metabolic processes stop together or that all enzymes are inactivated. Only the sum total of processes which lead to germination no longer operates. This was illustrated for cocoa beans by Holden (Table 3.3). For this reason all chemical or histochemical methods devised to test viability are only partially satisfactory. Such tests can only check for one definite reaction which may to some extent be correlated with the eventual ability of the seed to germinate. Most of these tests are based on the activity of certain oxidizing enzymes. The best correlation has been found to the activity of enzymes reacting with redox dyes, such as tetrazolium, but even here positive results do not

TABLE 3.3 — ENZYME ACTIVITY AND VIABILITY IN COCOA BEANS
The beans were allowed to ferment which raises their temperature
and eventually kills them
(After Holden, 1959)

	Enzyme activity as % of unfermented beans			
Time of fermentation	20	44	68	92 hr
Temp. °C	33	42	44	44
% germination	100%	0	0	—
Amylase	111	111	32	<5
β-Glucosidase	103	46	0	0
Catalase	160	20	0	—
Peroxidase	75	9	<5	<5
Polyphenol oxidase	73	17	15	11

always indicate 100 per cent germination of the seed population. General chemical changes in seed composition, as viability is lost, are discussed by Owen (1956).

Viability is retained for very long periods of time, especially in seeds having a hard seed coat, as in the Leguminosae where viability is often retained for several decades. The most extreme case of retention of viability is the case of *Nelumbo nucifera*, the Indian lotus. Seeds of this variety have been shown by radio carbon dating to be around a thousand years old. These seeds were found in the mud of a lake bed in Man-

TABLE 3.4 — CHANGES IN PERCENTAGE GERMINATION OF SEEDS DURING STORAGE
(compiled from various sources)

	Years of storage in which seeds germinate		
	70–100%	30–60%	less than 30%
Wheat	9	—	13
Rye	7	—	12
Barley	8	—	12
Oats	11	12	—
Melon	11		
Cucumber		9	
Spinach			5
Tobacco			11
Sunflower			9
Buckwheat			8
Alfalfa		11	
Clover (red)		4	
Clover (white)		2	
Peas	3		
Timothy grass		5	
Carrots	1	7	15
Eggplant	5	7	10
Lettuce	3	—	5
Onion	—	1	3
Pepper		1	5
Tomato	7	—	10
Flax		18	
Radish		10	

churia and germinated after their seed coats were broken. Even
if the radio carbon dating is not accepted, more conservative
estimates date the seeds as being about 250–400 years old. In
contrast all the claims of viability of grains found in the Egyp-
tian pyramids have been shown to be spurious.

The very long life span of many species of wild plants
must be contrasted with the relatively short one of many cultivated species. In these, from a practical point of view, a high
germination percentage is important and this is usually retained for relatively short periods. Some data are given in
Table 3.4.

II. External Factors Affecting Germination

In order that a seed can germinate, it must be placed in environmental conditions favourable to this process. Among the
conditions required are an adequate supply of water, a suitable
temperature and composition of the gases in the atmosphere,
as well as light for certain seeds. The requirement for these
conditions varies according to the species and variety and is
determined both by the conditions which prevailed during
seed formation and even more by hereditary factors. Frequently
it appears that there is some correlation between the environmental requirement for germination and the ecological conditions occurring in the habitat of the plant and the seeds. In the
following these various factors will be considered in detail.

1. *Water*

The first process which occurs during germination is the
uptake of water by the seed. This uptake is due to the process
of *imbibition*. The extent to which imbibition occurs is determined by three factors, the composition of the seed, the permeability of the seed coat or fruit to water, and the availability of water in liquid or gaseous form in the environment.
Imbibition is a physical process which is related to the proper-

ties of colloids. It is in no way related to the viability of the seeds and occurs equally in live seeds and in seeds which have been killed by heat or by some other means. During imbibition molecules of solvent enter the substance which is swelling, causing solvation of the colloid particles and, in addition, occupying the free capillary spaces and the intermicellar spaces of the colloid. The swelling of the colloid results in the production of considerable pressures, called *imbibition pressure*. This is usually measured and also defined as that pressure which must be applied to the system to prevent the colloid from swelling. The imbibition pressure developed by seeds may reach hundreds of atmospheres and in colloids such as agar or gelatine pressures of many hundred of atmospheres have been measured. The imbibition pressure is of great importance in the process of germination as it may lead to the breaking of the seed coat and also to some extent makes room in the soil for the developing seedling. The magnitude of the imbibition pressure is also an indication of the water retaining power of the seed and therefore determines the amount of water available for rehydrating the seed tissues during germination. In seeds we are dealing with the imbibition of water by hydrophilic colloids. Colloids are characterized by the size of the particles in the dispersion phase, and imbibition is a property of colloids which are in the form of a gel, i.e. where the colloidal particles constitute a more or less continuous micellar network, showing a certain amount of rigidity.

In hydrophilic gels the positive and negative charges are organized in a definite fashion, due to the presence of an electric double layer. This is made up of a very narrow fixed layer of solvent held to the surface of the solid and a second diffuse layer which extends into the bulk of the solvent. The difference of potential between the fixed layer and the freely mobile layer, is termed the *zeta potential*. Due to the zeta potential on the one hand and the dipole nature of the water molecules on the other hand, electrostatic forces play an important part during swelling of hydrophilic gels. Imbibition is also accompanied by the liberation of heat, particularly during the initial absorption of water. This is an indication that true compound formation occurs during the early stages of imbibition. The volume of the imbibing substance increases during imbibition, but the volume of the hydrated colloid is smaller than that of the sum of the solvent imbibed and the colloid, before imbibition occurred. This volume change is explained by the loss of a component of translational mobility of the system due to the sorption of water during the early stages of imbibition. (Glasstone, 1946, Jirgensons, 1958).

The gels occurring in nature, and in seeds in particular, are usually polyelectrolytes and contain a large number of ionic groups. The molecules themselves are of considerable molecular weight and are therefore not freely mobile. It is therefore possible to treat the imbibition of water into such polyelectrolytes according to the theories of a Donnan equilibrium. This treatment has led many authors to the view that imbibition as a whole should be treated as a special case of osmosis and that the driving forces involved are in fact the same as those concerned in osmosis. These studies consider that part of the swelling colloid, due to its immobility, acts as a semi-permeable membrane and the bulk of it as the osmotic system. (Haurowitz, 1950; Katchalski, 1954).

In seeds the chief component which imbibes water is the protein. However, other components also swell. The mucilages of various kinds will contribute to swelling, as will part of the cellulose and the pectic substances. Starch on the other hand does not add to the total swelling of the seeds, even when large amounts of starch are present. Starch only swells at very acid pH or after treatment with high temperatures, conditions which do not occur in nature. The swelling of seeds therefore to some extent reflects the storage materials present in the seeds (Table 3.5). Seed swelling depends on the pH of the solution but does not strictly follow the behaviour expected if only ampholytes were swelling. Proteins being "Zwitter-ions" show a minimum of imbibition at their isoelectric point, the imbibition rising with pH on either side of this point. Other colloids show a dependence of imbibition on pH

TABLE 3.5 — IMBIBITION BY VARIOUS SEEDS
Imbibition at 28°C expressed as percentage of the original
weight of the seeds. (After Levari, 1960)

Time in hours	Lettuce	Wheat	Sunflower	Vicia sativa (vetch)	Zea mays
1	170	114	124	112	111
2	185	120	137	143	116
4	197	127	147	168	—
6	201	133	153	181	124
10	213	140	154	182	—
16	225	—	—	—	136
24	237	151	—	—	137
32	252	155	—	—	—
40	254	—	—	—	—
48	270	161	—	—.	—

which can be related to their dissociation constant. For agar, maximal dissociation is near neutrality and maximal swelling also occurs around pH 7·0. Imbibition is dependent on temperature and proceeds more rapidly at higher temperatures.

Shull (1920) compared the imbibition of *Xanthium* seeds having a semi-permeable membrane and split peas without such a membrane. He noted that swelling of the seeds was essentially similar to that of colloids. The Q_{10} values of imbibition, in both cases, was between 1·5 and 1·8. Shull concluded that no chemical change was involved in the effect of temperature on imbibition and that it was not markedly affected by the presence of a semi-permeable membrane.

The effect of temperature on imbibition is probably complex. The viscosity of water decreases with increased temperature and its kinetic energy increases. The kinetic energy is directly proportional to the absolute temperature, while the molecular velocity varies as the square root of the absolute temperature.

Modern views on osmotically-induced flow of water assume that the bulk of the flow is hydrodynamic mass flow (Pappenheimer, 1953) through the pores of the membrane and

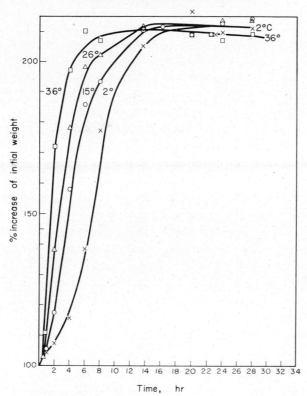

FIG. 3.1. Imbibition of heat-killed peas at different temperatures

×————————×	2°	△————————△	26°
□————————□	36°	○————————○	15°

not diffusion. As mentioned above, imbibition may be considered as a special case of osmosis. Therefore any effect of temperature on the structure of the colloid and the dimensions of its intermicellar spaces might affect the rate of imbibition. The final volume obtained by seeds imbibing at low temperature is greater than that resulting from the rapid imbibition at higher temperature (Fig. 3.1). However, experimental evidence shows that for seeds these differences are very slight.

The composition of the germination medium also determines the imbibition of seeds, as it determines the availability of water. This is of significance under natural conditions where the solution in which the seeds are found is not usually pure water. As the concentration of solutes in the solution increases, imbibition decreases. This is largely due to osmotic effects, as increasing osmotic pressures of the solvent due to increases in concentration of solutes decrease imbibition. In addition, however, a direct effect of the ions on the seed is also frequently observed. Toxic effects may be present, for example under very saline conditions.

The preceding discussion considered the problem of imbibition by the colloid materials in the seed. The entry of water into seeds is, however, determined in the first instance by the permeability of the seed coat or the fruit coat. Seeds which are surrounded by an impermeable seed coat will not swell even under otherwise favourable conditions. Impermeable seed coats are frequently found in *Leguminosae* as well as in other groups. The seed coat is usually a multi-layered membrane containing a number of layers or cells. Frequently it shows selective permeability toward certain substances. The impermeability of the seed coat, or its selective permeability, is frequently the cause of dormancy (see below). Various external

TABLE 3.6 — EFFECT OF EXTRACTING SEED COATS WITH HOT WATER OR HOT ALCOHOL ON THE WATER PERMEABILITY OF THE ISOLATED SEED COAT
(Compiled from data of Denny, 1917)

Seed	Solvent	Increase in permeability	Probable seed coat constituent restricting water permeability
Arachis hypogea	Hot water	170	Tannins
Arachis hypogea	Hot alcohol	80	Lipids (2% of seed coat)
Prunus amygdalis	Hot water	500	Pectic substances
Prunus amygdalis	Hot alcohol	350	Lipids
Citrus grandis	Hot water	0	Fatty substances
Citrus grandis	Hot alcohol	0	surrounded by thick pectinized walls
Cucurbita maxima	Hot water	0	Lipids
Cucurbita maxima	Hot alcohol	700	

factors can cause changes in the permeability of the seed coat. For example, heat-killed seeds often imbibe water more rapidly than the corresponding viable seeds, probably because the permeability of the seed coat is increased by the heat treatment. Denny (1917) showed that the seed coats of various species show different permeability to water. He was able to relate these differences to the composition of the seed coat and especially to lipoid components (Table 3.6). Permeability of the seed coat is generally greatest near the micropylar end of the seed, where it is almost invariably thinner than the rest of the seed coat.

2. Gases

Germination is a process related to living cells and requires an expenditure of energy by these cells. Energy-requiring processes in living cells are usually sustained by processes of oxidation, in the presence or absence of oxygen. These processes, respiration and fermentation, involve an exchange of gases, an output of carbon dioxide in both cases and also the uptake of oxygen in the case of respiration. Consequently seed germination is markedly affected by the composition of the ambient atmosphere. Most seeds germinate in air, i.e. in an atmosphere containing 20 per cent oxygen and a low percentage, $0 \cdot 03$ per cent, of carbon dioxide. However, many authors have shown that certain seeds respond to an increase in the oxygen tension above 20 per cent by increased germination, e.g. *Xanthium* and certain cereals.

Thornton(1934)showed that the germination of many cultivated seeds, if maintained at 20 per cent oxygen, was unaffected by increased concentration of carbon dioxide. Seeds of *Daucus carota* and *Rumex crispus* respond to increased oxygen concentrations in the dark by increased germination; 0 per cent germination at 20 per cent O_2, 3 per cent at 40 per cent O_2, and 24 per cent at 80 per cent O_2 (Gardner, 1921).

Most seeds will show lower germination if the oxygen tension is decreased appreciably below that normally present in the atmosphere. Although it is usually assumed that rice germinates well under anaerobic conditions, caused by flooding of the seeds, more recent work in China has thrown doubt on this. Tang, Wang and Chih (1959), and Chu and Tang (1959), showed that anaerobic conditions lead to the formation of abnormal seedlings and that these abnormalities are prevented by the presence of oxygen. In contrast, a number of seeds show increased germination as the oxygen content of the air is decreased below 20 per cent. The best established cases seem to be those of *Typha latifolia* and *Cynodon dactylon*, which germinate better in the presence of about 8 per cent oxygen than in air (Morinaga, 1926). It is important to note that quite different results are obtained if the dilution of the air is done with nitrogen or with hydrogen. For example, Morinaga found that while white clover, *Trifolium repens*, seeds give 52 per cent germination in air, they give only 47 per cent in air diluted with 60 per cent nitrogen, but give 70 per cent germination in the presence of air equally diluted with hydrogen. Clearly hydrogen has an effect on germination. In none of these cases is anything known about the oxygen tension within the seeds. Dormancy is often ascribed to the impermeability of the seed coat to certain gases. Such differential permeability of seed coats was demonstrated by Brown (1940) (see Chapter 4). The determining factor which may be involved in the effect of oxygen on germination will be considered in Chapter 4 on dormancy, as well as in Chapter 7.

The effect of carbon dioxide is usually the reverse of that of oxygen. Most seeds fail to germinate if the carbon dioxide tension is greatly increased, as shown for example by Kidd (1914) for *Hordeum vulgare* and *Brassica alba*. However, in some cases at least, there appears to be a minimal requirement for carbon dioxide in order that germination can occur. This seems to be so for *Atriplex halimus* and *Salsola*, as well as for lettuce. Other *Atriplex* species are resistant to high carbon dioxide concentrations, as shown by Beadle (1952) and as found for cultivated seeds, provided the O_2 concentration is kept constant. Very high carbon dioxide concentrations, which prevent germination, seem to have a favourable effect on the keeping of seeds. Kidd (1914) showed that the life-span of *Hevea brasiliensis* seeds was prolonged by sealing the seeds in an atmosphere containing 40–45 per cent CO_2. However an atmosphere of nitrogen was even more effective. For lettuce and onion seeds storage under carbon dioxide decreased the number of chromosomal aberrations which occur in the mitoses during the germination, subsequent to storage (Harrison and McLeish, 1954). Cases are known where increases in CO_2 concentration increase germination, e.g. *Phleum pratense* (Maier, 1933).

For carbon dioxide also, the determining factor probably is the internal concentration. This is determined by the permeability of the seed coat to carbon dioxide accumulating within the seed (see also dormancy).

3. *Temperature*

Different seeds have different temperature ranges within which they germinate. At very low temperatures and very high temperatures the germination of all seeds is prevented. The precise sensitivity is very different according to the species. A rise in temperature does not necessarily cause an increase in either the rate of germination or in its percentage. Germination as a whole is therefore not characterized by a simple temperature coefficient. This can be understood if it is appreciated that germination is a complex process and a change in temperature will affect each constituent step individually, so that the effect of temperature which is observed will merely reflect the overall resultant effect.

In studying the effect of temperature on germination one must distinguish between the resistance of dry seeds to various temperatures and the effect of temperature on the actual germination. Many dry seeds are fairly resistant to extreme temperatures. Thus it has been shown that the viability of seeds is not affected by placing them at the temperature of liquid air. High temperatures up to about 90°C for pro-

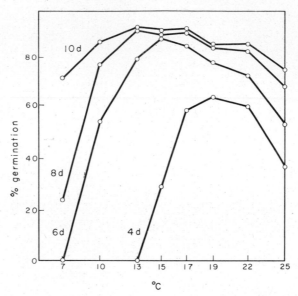

FIG. 3.2. Effect of temperature on the germination of oats.
(After Edwards, 1932 and Attenberg, 1928)
The percentage germination as determined at the different
temperatures after various periods of time, viz. 4, 6, 8
and 10 days.

longed periods of time are tolerated by many seeds such as
radish, turnips and poppy, as show by David (1936) who
studied oil-containing seeds. In other cases the temperature
tolerance is lower. Above 90°C the resistance of the seeds
is greatly lowered. There probably is some correlation between
the storage materials in the seed and their heat resistance, although seeds may still be viable after treatment at high temperature; the subsequent development of the seedling is often
adversely affected. Levitt, (1956)

   In the range of temperatures within which a certain seed
germinates there is usually an optimal temperature, below and
above which germination is delayed but not prevented. The
optimal temperature may be taken to be that at which the
highest percentage of germination is attained in the shortest
time. The minimal and maximal temperatures for germination

are the highest and lowest temperatures at which germination will occur. The minimal temperature is frequently ill defined because germination is so slow that the experiments are often terminated before germination could in fact have occurred. In many cases it is stated that the optimal temperature for germination shifts with time of germination. In these cases the term optimal is used ambiguously and in fact refers to the optimal temperature when germination is observed after some definite, arbitrary, time interval. If a different time interval is chosen, then a different temperature may be optimal according to this usage of the term (Fig. 3.2).

The temperature at which different seeds germinate and the range within which they germinate, is determined also by the source of the seeds, genetic differences within a given species, e.g. varietal differences, as well as age of the seeds. A few examples of ranges of temperature in which germination occurs are given in Table 3.7, but in view of the preceding discussion such data must be treated with great reserve.

TABLE 3.7 — TEMPERATURE RANGES IN WHICH GERMINATION OCCURS FOR DIFFERENT SEEDS. (Compiled from various sources).

Seeds	Temperature		
	Minimum	Optimum	Maximum
Zea mays	8–10	32–35	40–44
Oryza sativa	10–12	30–37	40–42
Triticum sativum	3–5	15–31	30–43
Hordeum sativum	3–5	19–27	30–40
Secale cereale	3–5	25–31	30–40
Avena sativa	3–5	25–31	30–40
Fagopyrum esculentum	3–5	25–31	35–45
Cucumis melo	16–19	30–40	45–50
Convolvulus arvensis	0·5–3	20–35	35–40
Lepidium draba	0·5–3	20–35	35–40
Solanum carolinense	20	20–35	35–40
Nicotiana tabacum (Florida cigar wrapper)	10	24	30
Delphinium (annual)	–	15	20–25

In contrast to those seeds which germinate readily if held at one specific temperature, instances are known where a periodic alternation of temperature is required for germination to occur, as in *Oeonothera biennis*, *Rumex crispus*, *Cynodon dactylon*, *Nicotiana tabacum*, *Holcus lanatus*, *Agrostis alba*, *Poa trivialis* and many others. The most usual cases of such alternations are diurnal ones, between a low and a high temperature. An examination of some of the results obtained with alternating temperatures suggests that what is determining germination is the actual alternation of temperature. The temperatures chosen, between which the seeds were alternated, appears to be of secondary importance, provided they are in a range within which the seeds can germinate and their viability is not affected. For example *Agrostis alba* seeds gave 69 per cent if alternated between 12° and 21°C, and 95 per cent for an alternation between 21° and 28°C or 21° and 35°C. At constant temperature the germination was 49 per cent at 12°C, 53 per cent at 21°C, 72 per cent at 28°C and 79 per cent at 35°C (Lehman and Aichele, 1931).

This effect of alternating temperatures on germination was analysed by Cohen (1958) for lettuce seeds, by measuring the actual temperature reached by the seeds. He concluded that neither the rate nor the duration of the temperature change is determining germination. He suggests that the actual change of temperature of the seeds themselves is the determining factor.

Stimulation by alternating temperatures has been variously ascribed to an effect of temperature on sequential reactions during germination or to mechanical changes occurring in the seed.

The results of Cohen indicate that neither of these interpretations is acceptable and that the change takes place in a macro-molecular structure in the seed, which in its original form prevents germination in some way.

The effect of temperature on germination is not independent of other factors. Thus instances of interdependence between temperature and light are known for celery, *Amaranthus* and other seeds, where light promotes germination at unfavourable high temperatures, but not at low ones.

Physalis franchetii seeds germinate well in the dark between 5 and 15°C. Between 15°C and 35°C they require light for their germination and the light requirement increases with the temperature (Baar, 1912). A very similar situation exists for lettuce seeds.

Temperature effects are also known in relation to after-ripening of seeds. This point will be considered later.

4. *Light*

Among cultivated plants there is very little evidence for light as a factor influencing germination. The seeds of most cultivated plants usually germinate equally well in the dark and in the light. In contrast, among other plants much variability in the behaviour toward light is observed. Seeds may be divided into those which germinate only in the dark, those which germinate only in continuous light, those which germinate after being given a brief illumination and those which are indifferent to the presence or absence of light during germination. Daily illuminations have also been shown to effect germination, the effects being similar to those of photoperiodism in flowering.

The importance of light as a factor in the germination of seeds has long been recognized. Already at the end of the nineteenth and the beginning of the twentieth century, a number of papers by Cieslar (1883), Gassner (1915) and by Lehman (1913) analysed some of these phenomena. It is possible that the light sensitivity of seeds has some relation to their germination in their natural habitat although such a view is contested by others (Niethammer, 1922). Under natural conditions seeds may be shed so as to fall on the soil or enter the soil or be covered by leaf litter, thus creating different conditions of light during germination. Among the species which have been investigated for their light response during germination at least half showed a light requirement. For example, Kinzel

(1926) lists hundreds of plant species which he divides into several categories. His first groups include those germinating at or above 20°C in the light (about 270 species) and in the dark (~114 species). Two further categories germinate in the light (~190 species) and in the dark (~81 species) after severe frost, and others showed similar behaviour after mild frost (~52 species in the light and ~32 species in the dark). A group of seeds which are indifferent to light or dark includes 33 species. A selection from his data is given in Table 3.8.

TABLE 3.8 — RESPONSE OF SEEDS OF DIFFERENT SPECIES TO LIGHT
(From data of Kinzel, 1926)

A — seeds where germination is favoured by light
B — seeds where germination is favoured by dark
C — seeds indifferent to light or dark

A	B	C
Adonis vernalis	Ailanthus glandulosa	Anemone nemorosa
Alisma plantago	Aloe variegata	
Bellis perennis	Cystus radiatus	Bryonia alba
Capparis spinosa	Delphinium elatum	Cystisus nigricans
Colchicum autumnale	Ephedera helvetica	
Erodium cicutarium	Evonymus japonica	Datura stramonium
Fagus silvatica	Forsythia suspensa	Hyacinthus candicans
Genista tinctoria	Gladiolus communis	
Helianthemum cha-		
maecistus	Hedera helix	Juncus tenagea
Iris pseudacorus	Linnaea borealis	Linaria cymbalaria
Juncus tenuis	Mirabilis jalapa	
Lactuca scariola	Nigella damascena	Origanum majorana
Magnolia grandiflora	Phacelia tanacetifolia	Pelargonium zonale
Nasturtium officinale	Ranunculus crenatus	Sorghum halepense
Oenothera biennis	Silene conica	Theobroma cacao
Panicum capillare	Tamus communis	Tragopogon pratensis
Resedea lutea	Tulipa gesneriana	Vesicaria viscosa
Salvia pratense	Yucca aloipholia	
Suaeda maritima		
Tamarix germanica		
Taraxacum officinale		
Veronica arvensis		

Such a classification of light requirement is probably an oversimplification, as light requirement varies during storage. In some species a light requirement only exists immediately after harvesting (e.g. in *Salvia pratensis*, *Saxifraga caespitosa* and *Epilobium angustifolia*), while in other species this effect persists at least for a year (e.g. *Epilobium parviflorum*, *Salvia verticillata* and *Apium graveolens*), while in yet other species it only develops during storage.

The question of light requirement has been the subject of detailed studies under laboratory conditions. These studies have shown, as is to be expected, that different spectral zones affect germination quite differently. The early work on the affect of light distinguished only between fairly wide spectral bands. These showed that light below 2900 Å inhibited germination in all seeds tested. Between 2900 Å and 4000 Å no clear-cut effects on germination were detected. In the visible range, 4000 Å–7000 Å, it was shown that light in the range 5600 Å–7000 Å and especially red light, usually promoted germination while blue light was said to inhibit. Flint and McAllister (1935 and 1937) determined these spectral ranges more accurately for lettuce seeds and showed that the most effective light in promoting germination is that having a wavelength of 6700 Å and that a germination inhibiting zone of the spectrum has its maximal activity at 7600 Å. This latter was far more effective than the inhibiting zone in the blue region of the spectrum.

Kincaid (1935) studied the effect of light on the germination of tobacco seeds. He found that as little as 0·01 seconds of sunlight was effective in stimulating germination and even moonlight could stimulate. Although the action spectrum was not studied in detail, a filter transmitting between 435–580 mμ with a maximum at 544 mμ was very effective. The seed coat had an absorption maximum at 510 mμ. Resuehr (1939) studied the germination of *Amaranthus caudatus;* the action spectrum showed three inhibitory zones, at about 450 mμ, between 475 and 490 mμ and between 700 and 750 mμ. Two promoting zones were observed at 640 mμ, where strong stimulation was obtained and at 675–680 mμ, where slight stimulation was noted. For the usually light-inhibited seeds of *Phacelia*, Resuehr observed light stimulation of germination at 640 mμ and five inhibitory zones, at 350 mμ, 450 mμ, 475–490 mμ, 680–690 mμ and above 1000 mμ. Orange light has been reported to stimulate germination of *Physalis franchetii* (Baar, 1912). It was more effective than blue-violet light.

Whether or not blue light does in fact inhibit germination or stimulate it, was disputed for many years, because it was contended that the light used by Flint and McAllister contained far-red or infra-red light as well as blue light. However, very recently both Wareing and Black (1957 and 1958) and Evenari, Neuman and Stein (1957) showed that blue light can in fact inhibit germination. The latter showed that under certain conditions blue light may also stimulate germination. Whether germination was stimulated or inhibited depended entirely on the exact period of illumination as related to the beginning of imbibition.

The sensitivity of seeds to light increases with time of imbibition. However, maximum sensitivity does not exactly coincide with the completion of imbibition (cf. Table 3.5 and Fig. 3.3). Moreover, seeds become sensitive to light long before they are fully imbibed. In fact, even storage of seeds at high relative humidities is sometimes sufficient to make them light-sensitive. If seeds are given a light stimulus while imbibed and then dried, the stimulatory effect is retained (Gassner, 1915; Kincaid, 1935). The precise time at which the seeds reach maximum sensitivity for any given species has been a matter of dispute. There have also been differences of opinion whether the sensitivity reaches a maximal value and then drops again or whether it remains at a high level. It appears that these differences can be related to the light intensity at which sensitivity was tested on the one hand, and to the time interval which was allowed to elapse between the illumination and the determination of germination percentage on the other hand. Figure 3.3 shows some of these effects for lettuce seeds.

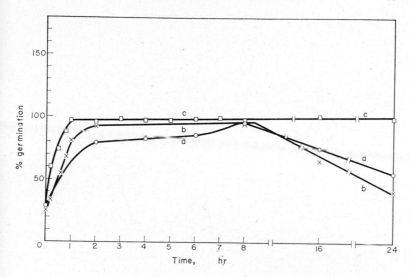

FIG. 3.3. Effect of length of imbibition period, prior to illumination, on light sensitivity of lettuce seeds, var. Grand Rapids, to red light at 26°C.

(a) o——o 30 sec light, total incubation time 48 hr
(Reconstructed from Evenari and Neuman, 1953)
(b) x——x 30 sec light, total incubation time 72 hr
(c) □——□ 2 min light, total incubation time 72 hr
(b), (c) after Poljakoff-Mayber and Lang, unpublished)

In tobacco seeds Kincaid (1935) found that high light intensities were effective after short periods of imbibition (several hours), while low light intensities were most effective after four days. After ten days of imbibition the seeds no longer responded to illumination.

The detailed work on the effect of light has been carried out on a relatively limited number of plant species, especially lettuce *(Lactuca sativa)*, *Lepidium virginicum*, *Nicotiana tabacum* and various *Amaranthus* species. For lettuce and *Lepidium* the precise spectral peaks for germination, stimulation and inhibition, by short illuminations, have been redetermined and earlier results essentially confirmed, showing stimulation at 6700 Å and inhibition at 7300 Å. An important new finding was the reversibility of both germination stimulation and

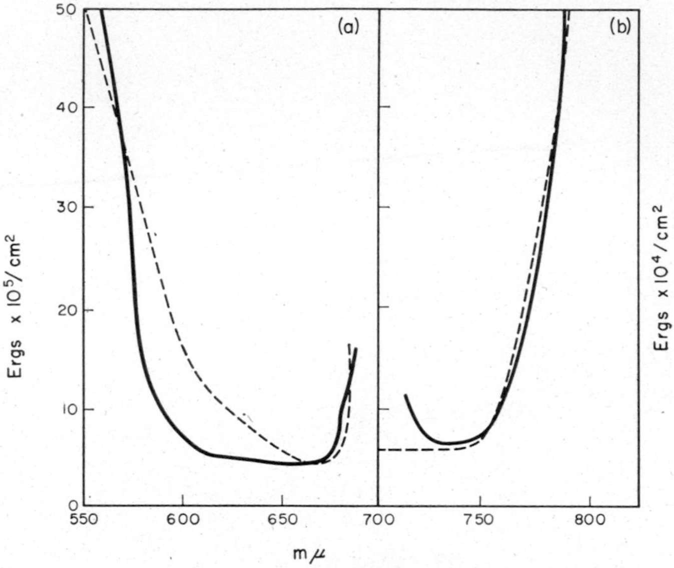

FIG. 3.4. Action spectrum for germination promotion (a) and inhibition (b) of *Lactuca sativa* and *Lepidium virginicum* seeds to 50% (After Toole *et al.*, 1956).

The curves show the amount of radiant energy at different wave lengths required to promote or inhibit germination to half its maximum value.

------- Lettuce, in (a) scale ×0·04, in (b) scale ×10
——— *Lepidium*

germination inhibition by alternating illuminations as shown by Borthwick *et al.* (1952). Lettuce seeds whose germination is stimulated by red light can be inhibited if they are subsequently illuminated with infrared (or far-red) light. Further illumination with red light will again induce germination. These effects are in fact very similar to, if not identical with, those also known for flowering, etiolation, unfolding of the plumular hook of bean seedlings as well as pigment formation in certain fruit and leaves.

All these phenomena show a very similar action spectrum. These action spectra have been deduced from observation of the biological response of the various tissues to illumination. The action spectra for the germination of lettuce and *Lepidium*

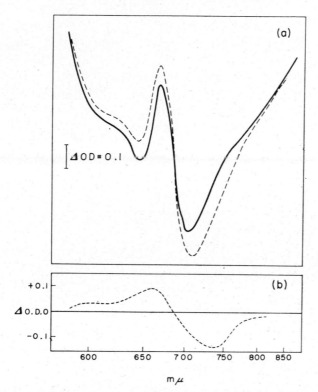

FIG. 3.5. Absorption spectra of maize coleoptiles following red or far-red irradiation (a) as well as the difference spectrum (b) (far-red irradiated spectrum minus red irradiated spectrum)

(Butler *et al.*, 1959)

———— Red irradiated

------- Far-red irradiated

seeds are shown in Fig. 3.4. Recently Butler *et al.* (1959) were able to show that there are changes in the absorption spectrum of an intact tissue, *Zea mays* coleoptiles, as a result of illumination with red or far-red light. This is illustrated in Fig. 3.5. These results are a great step forward in the study of the effect of light in processes such as germination, and bring clear evidence for the existence of a pigment system, which has long been postulated on the basis of the tissue response studies. Impure preparations of tissue extracts containing a photoreversible pigment, probably linked to a protein, have in fact been prepared by Butler *et al.*

The stimulation of germination by red light, and its inhibition by infra-red light, can be repeated many times and always the nature of the last illumination decides the germination response. As in all cases where light-effects are observed, so here also, the effect noted is dependent on both the intensity of the light, i.e. its energy, and the duration of the illumination. Provided that all the light is absorbed, then, within certain ranges, the response of germination will be a function of the product of light-intensity and duration of illumination, i.e. a function of the total energy of irradation. This applies both to germination stimulation and inhibition, as well as to all the other responses to red and far-red irradiation. In order to show the reversibility of red and far-red light effects, suitable intensities and durations must be selected. The range in which these observations have been made is for total irradiations of 2×10^4 ergs/cm^2 for promotion in lettuce and $1 \cdot 4 \times 10^6$ ergs/cm^2 for promotion in *Lepidium* and 6×10^5 and 3×10^4 ergs/cm^2 for inhibition respectively in these species (see also Fig. 3.4). The period of irradiation is between a few seconds and half an hour or more.

More detailed work by Hendricks and his co-workers (1959) has shown that these effects are rather more complicated than appears at first sight. Prolonged illumination of seeds with light of wavelength between 7000 and 8000 Å, with low light intensities (of the order of 10^5 ergs), or for shorter periods with higher light intensities ($> 10^7$ ergs), inhibited germination, for example in lettuce and *Nemophila insignis* seeds. The most effective zone was between 7000 and 7400 Å. This corresponds to the zone of reversible far-red inhibition. Inhibition in some cases also resulted from prolonged illumination at high energies in the region 6500–7000 Å. Kadman-Zahavi (1960) showed for *Amaranthus retroflexus* seeds, that prolonged far-red illumination abolishes the sensitivity of seeds to the effect of short illumination with red light. The sensitivity is restored when the seeds are kept for a prolonged period in the dark. However in lettuce seeds (variety, Great Lakes), prolonged high energy irradiation induced a response to short low energy illumination. This response was absent prior to prolonged illumination (Hendricks *et al.*, 1959). It seems that when the irradiations are given for longer periods of time, additional reactions occur. The nature of these is as yet quite unclear.

The inhibitory effect of prolonged illumination is tentatively explained by Hendricks *et al.* (1959). They suggest that continuous excitation of the pigment induces changes which inhibit germination. The final germination percentage seems to depend not only on the actual duration of the irradiations, but also on the time interval between stimulating and inhibiting irradiation and on the interval of time between the onset of imbibition and these irradiations. Kadman-Zahavi (1960) showed that in the interval between the red irradiation and the far-red irradiation, a temperature-dependent reaction occurs which presumably is biochemical in nature.

Numerous theories have been formulated to explain the effect of light on germination. They are all based on the justified assumption that a photochemical reaction is the primary step. In photochemical reactions only light absorbed is active and Einstein's law, that only entire quanta can induce a photochemical step, is observed. Consequently, as explained above, all theories concerning the effect of light on germination postulate the existence of an absorbing pigment.

The manner in which the pigment molecule reacts after light absorption is not entirely clear. Both Borthwicks *et al.* (1952) and Evenari and Stein (1953) have suggested theories to explain what happens to the pigment molecule. All these explanations are based on the view that following light absorption the pigment molecule is raised to an excited, possibly triplet, state. This excited molecule then in some way reacts with some other molecules, and initiates a series of reactions leading to germination. Whether this is a reaction by direct collision, or ion formation, or charge transfer reaction over larger distances up to 100 Å, is not clear. For fuller explanations of the mechanism of energy transfer the reader is referred to the *Discussions of the Faraday Society*, No. 27 (1959).

Another type of radiation which affects living organisms is short-wave irradiation such as γ-rays and X-rays. This type of radiation can also affect germination and development of seedlings. However, a consideration of the effects of this type of radiation as well as that of α and β-irradiation and the irradiation with other particles, is outside the scope of this book.

Strictly speaking, the effects of light discussed here are effects on the breaking of dormancy or its induction. Other light effects, perhaps more directly related to the phenomenon of dormancy, are the photoperiodic response of the germination of certain seeds as well as the complex interactions between light and other factors affecting germination. These will be dealt with in the following chapter.

BIBLIOGRAPHY

ATTENBERG, A. (1928) *Landw. Versuchsstation* **67,** 129.

BAAR, H. (1912) *S. Acad. Wiss. Math.-Naturn. K. L.* **121,** 667.

BEADLE, N. C. E. (1952) *Ecology* **33,** 49.

BEAL, W. J., as cited by SHULL, G. M. (1914) *The Plant World* **17,** 329.

BECQUEREL, P. (1932) *C. R. Acad. Sci. Paris* **194,** 2158

BECQUEREL, P. (1934) *C. R. Acad. Sci. Paris* **199,** 1662.

BORTHWICK, H A., HENDRICKS, S. B., PARKER, M. W., TOOLE, E. H. and TOOLE, V. K. (1952) *Proc. Nat. Acad. Sci., Wash.* **38,** 662.

BROWN, R. (1940) *Ann. Bot. N. S.* **4,** 379.

BUTLER, W. L., NORRIS, K. H., SIEGELMAN, H. W. and HENDRICKS, S. B. (1959) *Proc. Nat. Acad. Sci., U. S.* **45,** 1703.

CHING, T. M., PARKER, M. C. and HILL, D. D. (1959) *Agron. J.* **51,** 680.

CHU, C. and TANG, P. S. (1959) *Acta Bot. Sinica* **8,** 212.

CIESLAR, A. (1883) *Forsch. Gebiete Agrikultur Physik* **6,** 270.

COHEN, D. (1958) *Bull. Res. Council, Israel* **6D,** 111.

CROCKER, W. (1938) *Bot. Rev.* **4,** 235.

DAVID, R. L. (1936) *Influence des Températures Elevées sur la Vitalité des Grains Oleagineuses*, Imprimerie Universitaire, Aix en Provence.

DENNY, F. E. (1917) *Bot. Gaz.* **63,** 373.

DENNY, F. E. (1917) *Bot. Gaz.* **63,** 468.

Discussions of the Faraday Society No. 27 (1959) Energy transfer with Special Reference to Biological Systems. Aberdeen University Press, Scotland.

DUVEL, J. W. T. (1905) *U. S. D. A. Bureau of Plant Industry Bull.* 83.

EDWARDS, T. I. (1932) *Quart. Rev. Biol.* **7**, 428.

EVENARI, M. and NEUMAN, G. (1953) *Bull. Res. Council, Israel* **3**, 136.

EVENARI, M. and STEIN, G. (1953) *Experientia* **9**, 94.

EVENARI, M., NEUMAN, G. and STEIN, G. (1957) *Nature, Lond.* **180**, 609.

FLINT, H. L. and McALLISTER, E. D. (1935) *Smithsonian Miscellaneous Collection.* **94**, no. 5.

FLINT, H. L. and McALLISTER, E. D. (1937), *Smithsonian Miscellaneous Collection* **96** no. 2.

GARDNER, W. A. (1921) *Bot. Gaz.* **71**, 249.

GASSNER, G. (1915) *Z. Bot.* **7**, 609.

GLASSTONE, S. (1946) *Textbook of Physical Chemistry*, Macmillan.

GOSS, W. L. (1924) *J. Agr. Res.* **30**, 349.

GRIFFITHS, A. E. (1942) *Cornell Univ. Agr. Exp. Stat. Memoir*, 245.

HARRISON, B. J. and McLEISH, J. (1954) *Nature, Lond.* **173**, 593.

HAUROVITZ, F. (1950) *Chemistry and Biology of Proteins*, Academic Press, N. Y.

HENDRICKS, S. B., TOOLE, E. H., TOOLE, V. K. and BORTHWICK, H. A. (1959) *Bot. Gaz.* **121**, 1.

HOLDEN, M. (1959) *J. Sci. Food. and Agr.* **12**, 691.

JIRGENSONS, B. (1958) *Organic Colloids*, Elsevier, Amsterdam.

KADMAN-ZAHAVI, A. (1960) *Bull. Res. Council, Israel* **9 D**, 1.

KATCHALSKI, A. (1954) *Progr. in Biophysics*, **14**, 1.

KIDD, F. (1914) *Proc. Roy. Soc.* B **87**, 410.

KINCAID, R. R. (1935) *Tech. Bull.* **277**, Univ. Florida, Agr. Exp. St.

KINZEL, W. (1926) *Frost* und *Licht, Neue Tabellen*, Eugen Ulmer, Stuttgart.

LEHMANN, E. (1913) *Biochem. Z.* **50**, 388.

LEHMANN, E. and AICHELE, F. (1931) *Keimungsphysiologie der Gräser*, Verlag Ferdinand Enke, Stuttgart.

LEVARI R. (1960) Ph. D. thesis, Jerusalem.

LEVITT, J. (1956) *The Hardiness of Plants*, Academic Press, New York.

MAIER, W. (1933) *Jahrb. Wiss. Bot.* **78**, 1.

MORINAGA, T. (1926) *Amer. J. Bot.* **13**, 126.

MORINAGA, T. (1926) *Amer. J. Bot.* **13**, 159.

NIETHAMMER, A. (1957) *Biochem. Zschr.* **185**, 205.

OWEN, E. BIASUTTI (1956) The Storage of Seeds for Maintenance of Viability, *C. A. B., Bulletin* 43, Commonwealth Bureau of Pastures and Field Crops.

PAPPENHEIMER, J. R. (1953) *Physiol. Rev.* **33**, 387.

RESUEHR, B. (1939) *Planta* **30**, 471.

ROBERTS, E. H. (1960) *Ann. Bot.* **24**, 12.

SHULL, G. H. (1914) *The Plant World* **17**, 329.

SHULL, C. A. (1920) *Bot. Gaz.* **69**, 361.

TANG, P. S., WANG, F. C. and CHIH, F. C. (1959) *Acta Bot. Sinica* **8**, 199.

THORNTON, N. C. (1943–45) *Contr. Boyce Thompson Inst.* **13,** 355.

TOOLE, E. H. and BROWN, E. (1946) *J. Agr. Res.* **72,** 201.

TOOLE, E. H., TOOLE, V. K., BORTHWICK, H. A. and HENDRICKS, S. B. (1955) *Plant Phys.* **30,** 15.

TOOLE, E. H., HENDRICKS, S. B., BORTHWICK, H. A. and TOOLE V. K. (1956) *Ann. Rev. Plant Physiol.* **7,** 299.

TURNER, J. H. (1933) *Bull. Misc. Information* **6,** Royal Botanic Gdns. Kew, p. 257.

WAREING, P. F. and BLACK, M. (1957) *Nature, Lond.* **180,** 385.

WAREING, P. F. and BLACK, M. (1958) *Nature, Lond.* **181,** 1420.

WENT, F. W. and MUNZ, P. A. (1949) *El. Aliso* **2,** 63.

CHAPTER 4

DORMANCY, GERMINATION, INHIBITION AND STIMULATION

MANY seeds do not germinate when placed under conditions which are normally regarded as favourable to germination, namely an adequate water supply, a suitable temperature and the normal composition of the atmosphere. Such seeds can be shown to be viable, as they can be induced to germinate by various special treatments. Such seeds are said to be dormant or to be in a state of dormancy. Dormancy can be due to various causes. It may be due to the immaturity of the embryo, impermeability of the seed coat to water or to gases, prevention of embryo development due to mechanical causes, special requirements for temperature or light, or the presence of substances inhibiting germination.

In very many species of plants the seeds, when shed from the parent plant, will not germinate. Such seeds will germinate under natural conditions if they are kept for a certain period of time. These seeds are said to require a period of *after-ripening*.

After-ripening may be defined as any changes which occur in seeds during storage as a result of which germination is improved. This is the most generally used definition of this term. An alternative definition is also possible. This would define after-ripening as those processes which must occur in the embryo and which can occur only with time and which cannot be caused by any known means other than suitable storage of the seeds.

After-ripening often occurs during dry storage. In other cases storage of the seeds in the dry state does not cause after-ripening. The seeds must be stored in the imbibed state, usually at low temperatures, in order that they after-ripen. This is termed *stratification* and will be dealt with later. There is much diversity in the conditions under which after-ripening in "dry storage" occurs. The length of the storage period required is also variable. Thus some barley varieties after-ripen after a fortnight, while *Cyperus* after-ripens only over a period of seven years. After-ripening during dry storage is difficult to classify. In cereals germination is low at harvest and increases during storage. In lettuce, *Amaranthus* and *Rumex* the fresh seeds can germinate but the requirements for their germination are very specific. These special requirements tend to disappear during storage. For example, fresh lettuce seeds only germinate below 20°C but after storage germination occurs even at 30°C. The reverse case obtains in *Amaranthus retroflexus*. Fresh seeds only germinate above 30°C, while stored seeds germinate over a wider range down to 20°C. The light requirement in lettuce also disappears during prolonged dry storage. This type of response to storage should not strictly speaking be termed after-ripening, although this term is often applied to these phenomena.

The necessity for a period of after-ripening may be due to a number of factors. Various kinds of change may consequently occur during this process. In the case of an immature embryo further anatomical and morphological changes may occur. In other seeds chemical changes must occur in the seed before it can germinate. Frequently, germination of such seeds can be forced by suitable treatment although the resulting seedlings may be abnormal.

Immature embryos are known among many families of plants, although they are most commonly associated with plants which are saprophytic, parasitic or symbiotic. Such embryos may attain full maturity either during the actual process of germination or they may mature as a preliminary to germination. In either case the changes only occur if the seeds are kept under conditions favourable to germination, and differentiation between the two types is extremely difficult. Among the plants in which immature embryos occur are the Orchidaceae and Orobancheae, as well as some *Ranunculus* species. The period required for such embryos to reach maturity varies from a few days to several months. In all these, processes of differentiation at the anatomical and morphological level occur during the period of after-ripening. In other seeds, such as those of *Fraxinus*, the embryo although morphologically complete must still increase its size before germination can take place.

In contrast, in many seeds no visible anatomical or morphological changes occur in the embryo during after-ripening. In these it must be assumed that the process of after-ripening is the result of chemical or physical changes which occur within the seed or seed coat. The composition of the storage materials present in the seed may alter, the permeability of the seed coat may change, substances promoting germination may appear or inhibitory ones may disappear. In no case has it been possible to ascribe after-ripening to any one definite event.

I. SECONDARY DORMANCY

In contrast to those seeds which fail to germinate when shed, but germinate after a period of after-ripening, other seeds will germinate readily under favourable conditions. However, these seeds may lose their readiness to germinate. This phenomenon is called *secondary dormancy*. Secondary dormancy may develop spontaneously in seeds due to changes occurring in them, as in some species of *Taxus* and *Fraxinus*. These changes may be the reverse of those described as occurring during after-ripening. Sometimes secondary dormancy is induced if the seeds are given all the conditions required for germination except one. For instance, the failure to give light to light-requiring seeds, or illuminating light-inhibited seeds such as those of *Nigella* and *Phacelia*, induces dormancy. *Phacelia* seeds will not germinate if kept continuously in the light. When they are returned to, and kept in, the dark for long periods they will eventually germinate provided the period in the light has not been too prolonged. If the seeds are kept for very long in the light they are then unable to germinate in the dark under conditions under which they previously germinated (Mato, 1924).

Too high or too low temperatures for germination may also induce secondary dormancy, for example in *Ambrosia trifida* and *Lactuca sativa*. Lettuce seeds if kept imbibed in the dark at high temperatures will not germinate even if returned to lower temperatures, and no longer respond to light. Only much more drastic treatment such as a chilling or chemical treatment with gibberellin will induce germination.

Low oxygen tension will cause dormancy in *Xanthium* (Davis, 1930). High carbon dioxide tensions may cause secondary dormancy, e.g. in *Brassica alba* (Kidd, 1914). It may also be induced by chemical treatment of the seeds, e.g. the induction of light sensitivity in light-indifferent varieties of lettuce seeds by treatment with coumarin. This latter phenomenon is closely related to the problem of germination inhibitors and will be dealt with later.

II. POSSIBLE CAUSES OF DORMANCY

The possible causes of dormancy have already been listed. In the following these will be considered in greater detail. At the same time, the means by which seed dormancy may be broken will be discussed. Dormancy breaking is frequently of economic importance. The mechanism of artificial dormancy breaking and the natural process leading to the same effect are frequently similar.

1. *Permeability of Seed Coats*

A very widespread cause of seed dormancy is the presence of a hard seed coat. Such hard seed coats are met with in many plant families and usually can cause dormancy in one of three ways. A hard seed coat may be impermeable to water, impermeable to gases or it may mechanically constrain the embryo.

The impermeability of seed coats to water is most widespread in the Leguminosae. The seed coats of many members of this family are very hard, resistant to abrasion and covered with a wax-like layer. Such seed coats appear to be entirely impermeable to water. In some cases the entry of water into the seeds is controlled by a small opening in the seed coat, which is closed with a cork-like filling consisting of suberin —the strophiolar cleft with the strophiolar plug. Only if this plug is removed or loosened in some way, will water enter the seeds. This mechanism was first described by Hamly (1932) who also investigated various artificial ways of dormancy-breaking in such seeds. He was able to show that vigorous shaking of the seeds had the effect of loosening or removing the strophiolar plug, thus rendering the seeds permeable to water. This treatment is frequently called impaction and has been applied to seeds of *Melilotus alba*, *Trigonella arabica* and *Crotallaria aegyptiaca*. The majority of seeds showing a scrophiolar cleft belong to the Papillionaceae.

Other hard-coated seeds do not possess a strophiolar cleft. Such seeds become permeable to water only if the seed coat is abraded in some way. In nature, the seed coat may be broken down or punctured by mechanical abrasion, microbial attack, passage through the digestive tract of animals or exposure to alternating high and low temperatures which, by expanding and contracting the seed coat, crack it. Under laboratory conditions and in agriculture other means of rendering the seed coat permeable have been adopted. These are either shaking with some abrasive, to cause mechanical breakage, or chemical treatment. The chemical treatment is chiefly of two kinds: removal of the waxy layer of the seed coat by some suitable solvent such as alcohol, or treatment with acids (Table 4.1). The mechanism by which the latter act is far from clear. Possibly chemical decomposition of seed coat components is involved, which may be analogous to the breakdown processes occurring during microbial attack or passage through

TABLE 4.1 — EFFECT OF ACID TREATMENT ON GERMINATION
OF VARIOUS SEEDS
(From data of Wycherley, 1960; Khudairi, 1956)

	% Germination	
	Untreated	Treated with sulphuric acid
Calopogonum muconoides	30	65
Centrosema pubescens	30	70
Flemingia congesta	45	60
Pueraria phaseoloides	20	70
Purshia tridentata	1	84
Cercocarpus montanus	1	52
Atriplex canescens	11	16

the digestive tract. The treatment with alcohol is especially effective for members of the family Caesalpiniaceae. In all the cases studied, the various treatment could be shown directly to increase the water uptake of the seeds and the number of seeds which swelled (Table 4.2). The dormancy-breaking action of abrasion, alcohol or sulphuric acid could therefore be

TABLE 4.2 — CHANGES IN PERMEABILITY OF SEEDS TO WATER CAUSED
BY VARIOUS TREATMENTS
The change in permeability is expressed as the percentage
of seeds which swell
(From data of Barton and Crocker, 1948, and Koller, 1954)

Species	No treatment	Alcohol treatment	Impaction	Sulphuric acid	Mechanical abrasion
Trigonella arabica	5	4	100	—	96
Crotalaria aegyptiaca	0	20	80	95	95
Melilotus alba	1	0	86	—	100
Cassia artemisioides	2	57	3	—	100
Parkinsonia aculeata	2	100	8	—	100

directly related to an increase in the permeability of the seeds
to water. Nevertheless such vigorous treatment of seeds is
likely to induce other changes also, such as permeability to
gases, changes in sensitivity to light or temperature and possibly even destruction or removal of inhibitory substances.
Thus, although there is no doubt that increase in water permeability is an essential part of the dormancy-breaking action of
such treatments, it is by no means certain that it is the only
result of these treatments.

Frequently, seed coats are impermeable to gases despite
the fact that the seeds are permeable to water. This impermeability may be either toward carbon dioxide or oxygen or both.
That such differential permeability exists seems fairly well established, despite the apparently small differences in molecular
diameter of the substances involved. Only very few artificial
membranes are known which show such differential permeability.

The most frequently cited case of impermeability to oxygen
is provided by *Xanthium*. In *Xanthium* Crocker (1906) showed
that the fruit contains two seeds, an upper and a lower, which
differ in their ability to germinate. Later both Shull (1911 and
1914) and Thornton (1935) showed that they differ in their
requirement for external oxygen pressure for germination.
The upper seeds seem to require a much higher oxygen concentration than the lower ones, to give 100 per cent germination. The upper seed gives 100 per cent germination at 21°C
only in pure oxygen, while the lower one only requires 6 per
cent oxygen for full germination. The oxygen requirement of
both seeds is reduced by higher temperatures (Thornton,
1935). Intact seeds needed higher oxygen concentrations for
germination than excised embryos from both upper and lower
seeds, which give 100 per cent germination at 1·5 and 0·6 per
cent oxygen respectively. This indicates that the seed envelope
is impermeable to oxygen. It is not clear whether the impermeability of the seed envelope reduces the internal oxygen concentration to such an extent that oxidative respiratory mechanisms are reduced or whether some other mechanism is oper-

ative. The latter has been suggested by Wareing and Foda (1957) who indicated that the high oxygen requirement of the upper seeds is due to the presence of an inhibitor which has to be destroyed by oxidation before germination can occur.

Improved germination by increased oxygen tension is also shown by wild oats, indicating restricted permeability to oxygen. Spaeth (1932) reported that the nucellar membrane restricted the oxygen supply of seeds of *Tilia americana* and also advanced evidence that the testa is impermeable to moisture. In neither case is the mechanism clear.

Instances are known where seed coats are impermeable not only to oxygen but also to carbon dioxide. However, the significance of this is unclear. Brown (1940) has shown that the nucellar membrane of *Cucurbita pepo* shows a differential permeability to oxygen and to carbon dioxide. The isolated inner membrane is more permeable to carbon dioxide, $15 \cdot 5$ ml/cm^2 hr, than to oxygen, $4 \cdot 3$ ml/cm^2 hr, in *Cucurbita pepo*. Although the inner membrane is more permeable to gases than the outer one, it nevertheless is the inner membrane which in fact controls permeability. The permeability of the membrane to carbon dioxide is increased by treatment with chloroform or heat. This treatment is supposed to kill the living cells and thus to change the structure and permeability of the inner membrane. It did not affect the permeability to oxygen, indicating that the two gases followed different paths during diffusion through the membrane. The significance of this for the germination of the Cucurbitaceae is not clear. In *Xanthium*, carbon dioxide at high concentrations cannot induce dormancy of the intact seeds in the presence of oxygen. As little as 1 per cent oxygen was sufficient to reduce the effectiveness of carbon dioxide in inducing dormancy. The induction process was temperature-dependent (Thornthon, 1935).

Carbon dioxide does not invariably induce dormancy. In Chapter 3, a number of instances of improved germination have already been mentioned which are due to the presence of low concentrations of carbon dioxide. A more extreme case is that of the dormancy-breaking action of carbon dioxide on *Tri-*

folium subterraneum (Ballard, 1958). Carbon dioxide at concentrations between 0·3 and 4·5 per cent was effective. Carbon dioxide above 5 per cent was found to have an inhibitory effect. Treatment of the seed with activated carbon also stimulated germination and the results were consistent with the supposition that this treatment, too, resulted in a raising of the carbon dioxide concentration. These findings have been extended to other species of *Trifolium*, as well as to *Medicago* and *Trigonella* species. Carbon dioxide treatment caused breaking of dormancy in those cases where cold treatment was effective and even in some cases where cold treatment was ineffective, sometimes only after prolonged storage. Neither carbon dioxide nor cold broke the dormancy of the freshly-harvested seeds (Grant-Lipp and Ballard, 1959). The germination of *Phleum pratense* is stimulated by raised concentrations of CO_2 both in the light and in the dark, even after removal of the seed coat (Maier, 1933).

Axentjev (1930) studied the effect of seed coats on the germination of seeds of many species. In many seeds dormancy was partly caused by seed coat effects, probably due to the impermeability of the seed coat, and partly by a light requirement, and these factors are to some extent additive. Thus, *Cucumis melo* is light-inhibited, but its germination in the dark is further improved by removal of the seed coat (see Table 4.3).

2. *Temperature Requirements*

The effect of temperature on germination as such has already been discussed. An additional effect of temperature must be considered. Many seeds require an exposure to some definite temperature before they are placed at the temperature favourable to germination. Seeds are treated at either high or low temperatures which do not permit germination. They will germinate only on transfer to some other temperature. To respond to this treatment the seeds must be fully imbibed. Seeds requiring such temperature treatment must be consid-

TABLE 4.3 — EFFECT OF LIGHT AND SEED COATS ON GERMINATION OF VARIOUS SEEDS

Results given as % Germination

(Compiled from Axentjev, 1930)

Seed	Treatment	Temp.	5 Light	5 Dark	7 Light	7 Dark	8 Light	8 Dark	10 Light	10 Dark	12 Light	12 Dark	14 Light	14 Dark
Cucumis melo	Whole seed	17.5 to 20°C	0	41	4	65	5	72	20	78	30	87	68	100
	Without seed coat		0	89	15	96	25	98	48	100	60	100	80	100
	Cotyledons pierced, seed coat removed		19	84	32	98	36	100			42	100	48	100
	Root pierced with seed coat removed		6	72	18	93	18	97			23	100	27	100
Nigella arvensis	Whole seed	17.5 to 19°C	5	38	7	47	—	48						
	Seed coat pierced		2	18	6	48	8	51						
Oenothera biennis	Whole seed	14 to 20°C	0	0	0	—	0	—	0	0			3	—
	Seed coat pierced		1	1	13	5	16	7	24	9			34	12
Phacelia tanaceti-folia	Whole seed	7.5 to 10°C			22	92	22	—	53					
	Seed pierced				49	100	51	—						
	Whole seed	11.5 to 13°C			14	78								
	Seed pierced				69	100								

ered as being dormant and events occurring in them are akin to after-ripening in wet storage. The most commonly known and used procedure of exposure of the seeds to low temperatures under moist conditions is termed *stratification*. Evidently, during stratification changes take place within the seeds. Embryo growth has been noted during stratification. Such growth has been described in cherry seeds by Pollock and Olney (1959). These authors were able to show that during after-ripening in moist storage at 5°C the embryonic axis increased in cell number, dry weight and total length. An increased oxygen uptake was also noted in the embryonic axis and the leaf primordium, but not in the whole seed. On a cellular basis, an increase of oxygen uptake was especially marked in the embryonic axis. It appears that during after-ripening the energy supply to the embryo is increased. Investigations of the changes in nitrogen and phosphate content of cherry seeds during stratification were also carried out (Olney and Pollock, 1960). Nitrogen content rose in the embryonic axis and in the leaf primordium but the nitrogen content per cell remained constant. Phosphorus content, however, increased both on the basis of the total per organ and per cell. The

TABLE 4.4 — CHANGES OCCURRING IN CHERRY SEEDS DURING STRATIFICATION
(Compiled from graphs of Pollock and Olney, 1959: Olney and Pollock, 1960)

	Axis length, mm		No. of cells/axis		Dry Wt/axis (μg)		O_2 uptake of axis as % of non-after-ripened		Total N/axis (μg)		Total P/axis (μg)	
Temp (°C)	5°	25°	5°	25°	5°	25°	5°	25°	5°	25°	5°	25°
Initial	1·6	1·6	180	180	400	400	100	100	23	23	2·0	2·0
After 8 weeks	1·75	1·6	220	180	400	320	300	100	21	19	2·9	2·0
After 16 weeks	2·3	1·6	265	210	600	320	550	100	29	20	3·6	1·5

phosphorus and nitrogen which appears in the embryo presumably originate in the storage organs. Some of these results are shown in Table 4.4. All the changes were either absent or much less marked if the seeds were held at 25°C instead of at 5°C. Some of the changes observed in the phosphate content of the embryo are presumably related to the greater rate of respiration in this organ.

The metabolism of peach seeds during stratification has also been investigated (Flemion and de Silva, 1960). Marked changes in amino acid, organic acid, and phosphate composition were noted during after-ripening. However, it was impossible to establish a definite relationship between these changes and the actual ending of dormancy. A similar situation was found also with regard to growth substances. A later paper is of special interest because it showed that seedlings of peach could be obtained without chilling (Flemion and Prober, 1960). Seedlings were formed if the embryos were excised and the cotyledons removed. Thus, chilling is not absolutely essential to seedling establishment from these embryos. It is possible that the seedlings were formed from a lateral bud and not from the terminal bud. This seems more likely to Flemion and de Silva than that the cotyledons are actually inhibiting embryo growth, although this possibility is not excluded by them.

A number of enzymes have also been shown to change during stratification. Thus catalase and preoxidase were shown to increase in *Sorbus aucuparia*, *Rhodotypos kerrioides* and *Crategus* (Flemion, 1933; Eckerson, 1913). These changes in the enzymes may be the direct cause of emergence from dormancy but it seems much more likely that they are the secondary result of other changes in the seeds. For example, Barton (1934) was able to show a complete absence of correlation between increase in catalase activity during after-ripening and the completion of the after-ripening of *Tilia* seeds.

Many Rosaceous seeds contain cyanogenic glycosides such as amygdalin. These are often broken down during stratification, liberating hydrocyanic acid. There is some coincidence in time between the cessation of the liberation of HCN and the completion of after-ripening during stratification. However it is not known whether there is any causal relationship between the two processes.

In many cases no changes have as yet been observed during chilling treatment, but the evidence that before treatment the seeds failed to germinate and after treatment they did, indicates that internal changes nevertheless took place.

It is sometimes possible to force seeds requiring stratification to germinate by other means, e.g. by complete removal of the seed coats or by removal of the cotyledons (Flemion and Prober, 1960). Such experiments were carried out by workers at the Boyce Thompson Research Institute, particularly on Rosaceous seed. Seeds forced to germinate in this way do not form normal seedlings, the seedlings being either dwarfed or otherwise deformed (Barton and Crocker, 1948). However, cold treatment of the seedlings will again cause normal growth. Cold treatment apparently increases the growth capacity of the embryo and seedling. This could be the result of the disappearance of substances which inhibit growth or the formation of substances stimulating it. However, it could also be accounted for by other changes. There is some evidence to show that gibberellin-like compounds may be involved in some way, but this is by no means certain. Villiers and Wareing (1960) brought more direct evidence for the function of growth stimulators in these processes: in *Fraxinus* seeds a germination stimulator is formed during stratification (Fig. 4.1). It is assumed that dormancy is due to an inhibitor and that dormancy-breaking is the result of the counteraction of the inhibitor by the stimulator. The inhibitor content does not change during stratification. In this case gibberellins do not appear to be involved.

The conditions required for effective stratification are often similar to the natural conditions to which seeds are exposed. Many seeds are shed in autumn and then subjected to low temperatures during the winter months while they are exposed to moist conditions under leaf-litter or in the soil.

Although this would lead one to expect that stratification requirements are similar in many plants they in fact differ greatly. This difference exists both with regard to the length of the cold treatment and also with regard to the actual temperature during the treatment. Even in one family, e.g. the Rosaceae, and within one genus, marked differences exist. For example Barton and Crocker (1948) cite experiments to

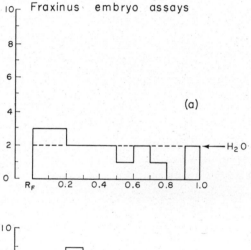

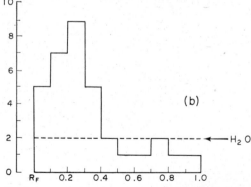

FIG. 4.1. The development of a growth stimulator in *Fraxinus* seeds during stratification (Villiers and Wareing, 1960).

Extracts were prepared from non-stratified seeds (a) and stratified seeds (b). The extracts were separated by paper chromatography and the various zones assayed with the *Fraxinus* growth test. The ordinate shows the number of embryos which germinated.

show that while *Rosa multiflora* has a requirement of two months at 5–10°C, *Rosa rubiginosa* requires six months stratification at 5°C and will not respond to treatment at either 10°C or 0·5°C. However, such a precise temperature requirement is rare and usually stratification can be carried out within a fairly wide temperature range. The best known cases of seeds requiring low temperature treatment for germination are found among the Rosaceae and among various conifers. However, in other families too, such requirements are known, as for instance in *Juglans nigra* (Juglandaceae), *Asters* (Compositae), *Adonis* and *Anemone* (Ranunculaceae), *Iris* (Iridaceae) as well as many others, including aquatic plants. In the latter the temperature treatment must be given while the seeds are actually immersed in water. A fuller list of plants requiring stratification, giving details of time and temperatures used, is given by Crocker and Barton (1953).

There is very little evidence in the literature to indicate that high temperatures in themselves break dormancy. Cases are known, as already mentioned, where high temperatures cause a change in the structure of the seed coats, thus causing a change in permeability. In contrast, many instances are known where there is interaction between the effects of low temperatures and somewhat raised ones. Probably a differential temperature requirement for growth of different parts of the embryo is here involved.

Crocker and Barton (1953) state that the seeds of *Paeonia suffruticosa* germinate by root growth, but the shoot fails to develop. If the germinated seeds are kept at normal temperatures, no further development occurs. However, if they are placed at a low temperature (1–10°C) for two or three months and kept moist, the bud of the epicotyl shoot develops normally when the seedlings are again transferred to normal temperatures. This phenomenon is called epicotyl dormancy by these authors. Similar phenomena occur in some species of *Lilium*.

Other cases are known where brief exposure to slightly elevated temperatures promotes subsequent germination at lower temperatures after exposure to light, for example in *Poa pratensis* and *Lepidium virginicum*. The mechanism of the effect of the raised temperatures is in no way clear.

In many seeds germination is promoted by alternating temperature changes which may be either diurnal or seasonal. Diurnal temperature changes have already been discussed in Chapter 3. (See also Table 4.5). A clear-cut effect is observable for *Nicotiana* in the dark when alternated between 20 and 30°C diurnally. Smaller effects are observable for some of the other seeds when the germination at 20 or 30°C is compared with that when they are alternated between 20 and 30°C. Seasonal changes of temperatures may effect germination by affecting the actual development of the embryo or in other ways. The practice of stratification already mentioned is, in fact, analogous to one kind of seasonal change of temperature. Other combinations of temperature alternations are met with, depending on the type of seed. Stratification is characterized by

TABLE 4.5—EFFECT OF LIGHT AND TEMPERATURE ON THE GERMINATION OF VARIOUS SEEDS
(Compiled from data of Toole *et al.*, 1955, and Koller, 1954)

Seed	Light condition	% germination at temperature stated					
		15°	20°	25°	30°	35°	20–30° C
Brassica juncea	Red light	90	48	18	2	–	80
	Dark	53	20	6	0	–	34
Lepidium virginicum	Red light	21	32	33	0	0	29
	Dark	0	0	0	0	0	1
Nicotiana tabacum	Red light	94	96	94	84	8	97
	Dark	2	5	6	0	0	97
Zygophyllum dumosum	White light	77	84	72	12	–	–
	Dark	80	82	82	16	–	–
Calligonum comosum	White light	0	4	8	0	–	–
	Dark	0	64	80	12	–	–
Juncus maritimus	White light	62	74	—	70	–	–
	Dark	0	0	—	0	–	–

exposure to a low temperature followed by a high one. In the case of the *Paeonia* seeds the reverse requirement has been shown.

In *Convallaria majalis* full development of the seedling is extremely complicated. Root protrusion occurs at 25°C, but is greatly promoted if the seeds are first exposed to a low temperature, followed by a raised temperature to permit root growth. The first leaf enters dormancy after it has broken through the cotyledonous sheath and requires at this stage low temperature treatment at 5°C, for its further development. Treatment at an earlier stage is ineffective. Further growth of the first leaf proceeds at normal temperature. The second leaf apparently also requires low temperature treatment for its development. Under normal conditions, in the natural habitat, such a cycle is not attained in less than nine months and may take up to fourteen months (Barton and Schroeder, 1942).

The literature on germination is full of examples where the temperature requirement is altered by exposure of the seeds to high, low or alternating temperatures. It seems clear that the changes occurring are extremely complex and affect primarily the rate of germination at the various temperatures. We have previously defined the optimal temperature of germination as involving both a rate factor and the actual germination percentage attained. Treatments with high, low or alternating temperatures can apparently widen the temperature range in which germination occurs; whether the optimal temperature is changed is not clear. This is illustrated by the behaviour of a number of seeds. *Betula lenta* seeds normally germinate only around 30°C. If the seeds are exposed to low temperature during stratification, the temperature at which the seeds can germinate is lowered. After prolonged stratification the temperature at which germination still occurs, is reduced to 0°C. *Festuca* seeds will not germinate at 30°C when freshly harvested, but storage at 20–30°C for one year will cause them to germinate at 30°C. However, storage at low temperature does not have this effect. The seeds still fail to germinate at 30°C. Another instance is provided by annual *Delphinium* seeds, which will germinate at temperatures up to 30°C, following stratification at 5–10°C or even 15°C for two months, but not without this treatment.

3. Light Requirements

The fact that light can act as a dormancy-breaking agent has already been mentioned. The question of the mechanism of light stimulation of germination has also been discussed in Chapter 3. In addition to those seeds which require light for their germination, there are many species whose germination is inhibited by light (see Tables 4.3 and 4.5 as well as Table 3.8).

Light does not only affect the absolute germination percentage but also the rate of germination. Thus in *Agrostis* seeds the final germination percentage is directly related to the light intensity. However, at high light intensities it takes much longer to reach maximal germination than at low ones (Leggatt, 1946).

In all these cases, no matter whether light inhibits or promotes, there is a complex interaction between light and other external conditions as well as with the age of the seeds. For example seeds of *Salvia pratensis, S. sylvestris, Epilobium angustifolium, Echium vulgare* and others lose their light sensitivity very shortly after harvest. In *Salvia verticillata, Epilobium parviflorum* and *Rumex acetosella* sensitivity is retained at least for a year. In other cases sensitivity is retained for much longer periods (Niethammer, 1927).

An instance of the complexity of the situation is provided by the case of lettuce seeds, variety Grand Rapids, which are sensitive to light. These seeds, immediately after harvest, hardly germinate at all in the dark at 26°C. After a certain period of storage they germinate in the dark at 18°C, but require a light stimulus for germination at 26°C. Peeling the seeds, including removal of the endosperm, or merely puncturing the endosperm, which encloses the embryo, is sufficient to abolish their light requirement at 26°C. This light requirement slowly decreases as the length of the storage period increases and eventually, after several years of storage, germination in the dark at 26°C reaches 60–80 per cent. It is possible to induce a light requirement in lettuce seeds which do not normally show it by treating the seeds with coumarin or by treatment with high temperatures or solutions of high osmotic pressure.

Such an induction of a light requirement was demonstrated by Toole (1959) for lettuce seeds, variety Great Lakes. These seeds germinate 95 per cent at 20°C in the dark. If kept for 4 days at 35°C they will only germinate 11 per cent in the dark at 20°C. However, irradiation with red light before returning them to 20°C restored germination to 95 per cent.

A more complicated interaction between light and temperature is shown by lettuce seeds, variety Grand Rapids. Seeds that gave about 30 per cent germination at 26°C, if kept for two days at 30°C and then returned to 26°C, only germinate about 1–4 per cent. Irradiation of such seeds with red light, following the treatment at 30°C, raises the germination percentage to 12–17 per cent, if the seeds are again returned to 26°C. Controls not given treatment at 30°C will germinate 95 per cent after this irradiation.

This interaction of light response of seeds with temperature was also clearly shown by Toole (1959) for seeds of *Lepidium virginicum*. Such seeds will not germinate in the dark if kept at 15°C or 25°C, or if transferred from 15 to 25°C or vice versa. If illuminated with red light, two days after sowing they will germinate to about 30 – 40 per cent, if kept throughout at either 15 or 25°C or on transfer from 25 to 15°C. However, the reverse transfer, from 15 to 25°C, when the light is given after the first two days, causes full germination. From this work and also from the examples quoted in Chapter 3, it is quite clear that the photochemical reaction in germination, no matter what its precise nature, is very closely linked to other, chemical, reactions, which will determine whether and to what extent there will be a response to illumination.

In many seeds the light effect is related to the presence of the seed coat. In lettuce seeds removal of the endosperm removes the light requirement. As will be seen from Table 4.3, the inhibition caused by light in the case of *Cucumis melo* is reduced by removal of the seed coat. In *Agropyron smithii*, removal of the seed coat does not abolish the inhibitory effect of light on germination. An unusual example is provided by *Oenothera biennis*. The intact seeds of this plant do not germinate in light or dark. However, if the seed coat is pierced, the seeds will germinate in light. Germination in the dark is only slightly promoted.

An entirely different light response is that of seeds whose germination is affected not by short illumination but by alternations between dark and light, i.e. by photoperiod. Photoperiodic response of seeds were first shown by Isikawa (1954 and 1955). He suggests that both light-stimulated and light-inhibited seeds may be considered to show short day behaviour. Thus, in the light-requiring seeds of *Patrina* and *Epilobium*, which germinate under illumination of from 2 to 21 hours, germination is promoted by interruption of the light period by darkness. Light-inhibited seeds such as *Nigella damascena* and *Silene armeria* germinate, if illuminated for one minute or 3 hours daily respectively, with low light intensities. The only instances quoted by Isikawa indicating a long day requirement are seeds of *Leptandra* and *Spiraea*.

A true long day requirement for germination appears to exist in *Begonia* seeds. The seeds germinate if given at least three cycles of long days, the critical day length being 8 hours. A break in the dark period, if illumination is less than 8 hours, is also effective in causing germination. Furthermore, the response to light is increased in the presence of gibberellin, the critical day length being greatly shortened (Nagao *et al.*, 1959).

Other more complex examples are provided by the work on *Betula* (Black and Wareing, 1955), on *Tsuga canadensis* (Stearns and Olson, 1958) and on *Escholtzia argyi* (Isikawa and Ishikawa, 1960). *Betula* seeds will at 15°C germinate only under long day conditions, eight cycles being required to induce germination. However at 20–25°C germination will occur following a single exposure to light for 8–12 hours. *Escholtzia* seeds do not germinate either in the light or the dark at constant temperatures. The seeds will however germinate if the light treatment is combined with a thermoperiodic treatment. This is illustrated by Table 4.6. From this it appears that highest germination is obtained if a 6 hour photoperiod at 25°C is followed by a 18 hour thermoperiod at 5°C. The results shown in Table 4.6 are difficult to analyse as both photo- and thermoperiod were varied simultaneously. It is never clear to what change increased or decreased germination must be ascribed.

Citrullus colocynthis seeds provide another instance of photoperiodic response in germination. Seeds of this plant germinate in the dark at 20°C. However exposure to long days, i.e. 12 hours of light per day, inhibits their germination. Short day treatment, 8 hours of illumination, does not inhibit. These phenomena thus show a marked similarity to photoperiodism in flowering. However, in no case has a requirement for a criti-

TABLE 4.6 — PERCENTAGE OF GERMINATION OF SEEDS OF
ESCHOLTZIA WITH DAILY PHOTOPERIODIC AND
THERMOPERIODIC TREATMENTS
(After Isikawa and Ishikawa, 1960)
Each temperature and light treatment was repeated eight
times; the initial treatment was done in light

Temperature and period of light exposure	Temperature and period of darkness	Germination (%)
15°, 6 hr	5°, 18 hr	60
15°, 18 hr	5°, 6 hr	24
25°, 6 hr	5°, 18 hr	80
25°, 18 hr	5°, 6 hr	25
25°, 6 hr	15°, 18 hr	40
25°, 18 hr	15°, 6 hr	10
35°, 6 hr	5°, 18 hr	63
35°, 18 hr	5°, 6 hr	28
35°, 6 hr	15°, 18 hr	66
35°, 18 hr	15°, 6 hr	0

cal dark period been demonstrated in germination. Moreover, in the photoperiodic response in germination, the induction phenomenon is often absent. Usually only continued treatment with the stimulating photoperiod will ensure germination, or continued treatment with an inhibiting period prevent it. As soon as the treatment is discontinued, the seeds will respond to the conditions under which they are placed. One of the exceptional cases in this respect is the behaviour of *Betula* seeds at 15°C, which show some kind of induction phenomena. What seems to be a clear case of induction exists in *Begonia* seeds (Nagao *et al.*, 1959). In these, germination occurs if the seeds are given six cycles of a photoperiod of 9 or more hours, followed by 7 days of darkness.

4. *Germination Inhibitors*

A very large number of substances can inhibit germination. All those compounds which are toxic generally to living organism will also, at toxic concentrations, prevent germination, simply by killing the seed. Far more interesting is the action of those substances which prevent germination without effecting the seeds irreversibly. The simplest case, and one perhaps most frequently met with in nature, is that of osmotic inhibition. It is possible to prevent the germination of seeds simply by placing themi n a solution of high osmotic pressure. When the seeds are removed from such an environment and placed in water, they can then germinate. Such a situation appears to exist in fruits and in the case of seeds of plants from saline habitats. The substances responsible for the high osmotic pressure may be sugars, inorganic salts such as sodium chloride, or other substances. The threshold of osmotic pressure at which germination is prevented, differs widely for different seeds. A few examples of these differences are shown in Fig. 4.2. In the laboratory, a convenient compound for obtaining osmotic inhibition is mannitol. The use of mannitol frequently gives results somewhat different from those obtained with sodium chloride. This is clearly shown in Fig. 4.3, and may be ascribed to the ionic toxicity of the salt. It appears that the germination percentage obtained is not strictly a function of the water uptake by the seeds. As long as a certain minimal amount of water is taken up, germination can occur, provided that no toxic effects of the osmotic medium are present.

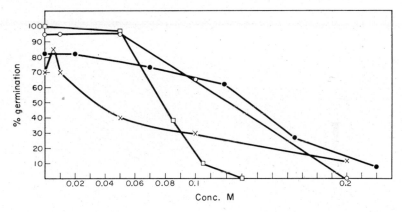

FIG. 4.2. The effect of sodium chloride concentration on the germination of various seeds.

(Compiled from data of Uhvits, 1946; Lerner *et al.*, 1959; and Poljakoff-Mayber, unpublished)

×——× *Atriplex halimus*	●——● Alfalfa var. Arizona Chilean
o——o Tomato, var. Marmand	□——□ Lettuce

Germination percentages determined after 8 days for tomatoes, 4 days for *Atriplex* and 2 days for lettuce and alfalfa.

Another type of inhibition is that caused by substances which are known to interfere with certain metabolic pathways. As germination is closely associated with active metabolism, all compounds which interfere with normal metabolism are likely to inhibit germination. An example of this type of interference is provided by various respiratory inhibitors. Compounds such as cyanide, dinitrophenol, azide, fluoride, hydroxylamine and others, all inhibit germination at concentrations similar to, but not always identical with, those inhibiting the metabolic processes. These substances probably act on germination as a result of their effect on metabolism. The action of such compounds is not necessarily that of induction of dormancy. It is, however, interesting to note that Mancinelli (1958) claims that ethionine and hydroxylamine appear to increase the light requirement of lettuce seeds, or at any rate

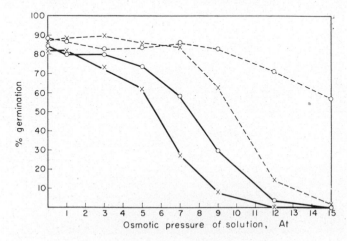

FIG. 4.3. The effect of osmotic pressure and ionic toxity on the germination of alfalfa seeds

(From data of Uhvits, 1946)

o——o Germinated in Mannitol for 2 days
o----o Germinated in Mannitol for 10 days
x——x Germinated in NaCl   for 2 days
x----x Germinated in NaCl   for 10 days

that light is able partially to reverse the inhibition caused by these substances. While the light reversal occurred both at 22°C and 26°C for ethionine, in the case of hydroxylamine the effect seemed more marked at 26°C.

Herbicides of various kinds inhibit germination to a greater or lesser extent. Many of the commonly used substances, such as 2,4-D, affect germination at comparatively low concentrations (Table 4.7). The more effective of such herbicides can and have been used in order to prevent the germination of weed seeds in agricultural crops. No selective compound

TABLE 4.7 — EFFECT OF 2,4-D AND COUMARIN ON GERMINATION OF
VARIOUS SEEDS
The results are given as the concentration of the compound
causing 50 per cent inhibition of germination
(Compiled from data of Audus and Quaster, 1947; Mayer and
Evenari, 1953; Isikawa, 1955; Libbert, 1957: Misra and Patnaik, 1959)

Seed	Molar concentration	
	2,4-D	Coumarin
Cress		0.75×10^{-4}
Radish		0.69×10^{-4}
Mustard		$>1.3 \times 10^{-4}$
Carrot	5.0×10^{-5}	0.35×10^{-4}
Beetroot		$>1.3 \times 10^{-4}$
Taraxacum		0.41×10^{-4}
Cabbage		0.27×10^{-4}
Onion		$>1.3 \times 10^{-4}$
Lettuce var. Grand Rapids	3.0×10^{-5}	1.5×10^{-4}
Lettuce var. Progress	6.0×10^{-5}	5.0×10^{-4}
Setaria		5.0×10^{-4}
Linum		3.3×10^{-5}
Rice		6.6×10^{-3}
Wheat	1.1×10^{-3}	$1.5-3.0 \times 10^{-3}$

that will satisfactorily distinguish between crop and weed
seeds appears to have been developed yet. A very frequent use
of herbicides is as pre-emergence weed killers. In these cases
the herbicide is applied in order to kill the seedling immediately after germination and before the main crop germinated.
In these cases the herbicides are not in fact acting as germination inhibitors.

Phenolic compounds of various kinds inhibit germination. Some of the selective herbicides, such as the substituted
phenols and cresols, inhibit germination due to their general
phytotoxicity. In addition to these compounds, many other
phenolic substances also inhibit germination. A list of a few
of these and the concentration at which they inhibit the germination of lettuce seeds, is given in Table 4.8. Many other phenolic substances at similar concentration also affect the germi-

nation of a variety of seeds. Because of the widespread occurrence and distribution of phenolic compounds in plants and fruits, it has been suggested that these substances might act as natural germination inhibitors (van Sumere, 1960). This problem will be returned to later.

TABLE 4.8 — THE INHIBITION OF GERMINATION OF LETTUCE SEEDS, PROGRESS VARIETY, BY A NUMBER OF PHENOLIC COMPOUNDS (Mayer and Evenari, 1952 and 1953)

Compounds	Concentration causing 50% inhibition
Catechol	$10^{-2}M$
Resorcinol	$5 \times 10^{-3}M$
Salicylic Acid	$1.5 \times 10^{-3}M$
Gallic Acid	$5 \times 10^{-3}M$
Ferulic Acid	$5 \times 10^{-3}M$
Caffeic Acid	$> 10^{-2}M$
Coumaric Acid	$5 \times 10^{-3}M$
Pyrogallol	$10^{-2}M$

All compounds mentioned so far are inhibitory but cannot be regarded as being dormancy-inducing. The term dormancy-induction is used to imply that the seed can again be made to germinate by one or other of the treatments already mentioned earlier, which break natural dormancy. Coumarin has been widely used as a germination inhibitor in the laboratory. As already mentioned, coumarin can induce light sensitivity in varieties of lettuce seeds not requiring light for germination, as first shown by Nutile (1945).

Coumarin and its derivatives are of fairly widespread occurrence in nature. Coumarin itself (Fig. 4.4) is characterized by an aromatic ring and an unsaturated lactone structure. The effect of changes in structure of the coumarin molecule on its germination inhibiting activity has been investigated in some detail both for wheat and lettuce seeds. This work showed that there was no single essential group in the coumarin molecule which was the cause of its inhibiting action. Reduction of the unsaturated lactone ring or substitution by most substituents such as hydroxyl, methyl, nitro, chloro and others in the ring system, all reduced the inhibitory activity of coumarin. A certain variation in the response of the two species tested were observed, which, however, did not materially alter the general picture (Mayer and Evenari, 1952). Opening of the lactone rings also reduced the inhibitory action of coumarin considerably.

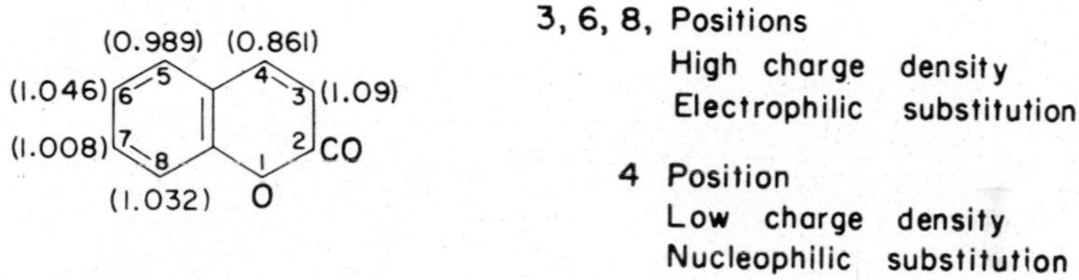

(0.989) (0.861)

(1.046) 6 5   4 3 (1.09)

(1.008) 7 8   1 2 CO

(1.032) O

3, 6, 8, Positions
High charge density
Electrophilic substitution

4 Position
Low charge density
Nucleophilic substitution

Fig. 4.4. Structure of coumarin with ring numbering and showing charge density.

The inhibitory action of coumarin has been studied on a wide variety of seeds and it has usually been found to inhibit germination. A few isolated instances of stimulation of germination by coumarin at very low concentrations are, however, known. The inhibitory concentration is different for different species and even differs in different varieties of the same species (Table 4.7). From these figures it will be seen that coumarin is extremely active as a germination inhibitor on a very wide variety of plant seeds. Because of its widespread distribution in plants and due to its strong inhibitory action, coumarin is considered to be one of the substances which may function as a natural germination inhibitor. This has been frequently proposed, but the occurrence of coumarin itself in seeds at inhibitory concentrations has been proved only in one instance, in the case of *Trigonella arabica* (Lerner *et al.*, 1959). However, coumarin derivatives such as the glycosides of the lactone, or substituted coumarins have been shown to occur in many fruits. For this reason the supposition that substances of this nature do function as germination inhibitors seems reasonable.

Auxins in high concentrations generally inhibit germination. In some cases gibberellin has also been shown to inhibit germination (Fujii *et al.*, 1960).

Wiesner first suggested at the end of the nineteenth century that seeds of *Viscum* contain a germination inhibitor which prevents their germination within the fruit. This idea was further developed by other workers including Molisch. It appears that Oppenheimer (1922) was among the first to study this problem experimentally. He tested among other things the juice of tomatoes, to determine whether they contain a germination inhibitor and concluded that they did indeed contain such an inhibitory substance. This conclusion has since been disputed. Some authors claim that the inhibitory action is solely due to osmotic effects while others claim to have isolated the inhibiting substance. Thus Akkerman and Veldstra (1947) claim to have shown that caffeic and ferulic acids are the compounds responsible for the inhibition of germination of seeds within the tomato fruit. Despite this claim it must be pointed out that these authors failed to show that caffeic and ferulic acids occur in the tomato at concentrations sufficiently high to account for the inhibition of germination of the seeds in the mature fruit. It seems likely that the explanation of the failure of tomato seeds to germinate in the fruit can be ascribed to the combined action of osmotic inhibition together with a specific or non-specific germination inhibitor. Such interaction between the osmotic effect and the effect of an inhibitor has been shown to exist for example in the case of lettuce seeds (Lerner *et al.*, 1959). It seems very probable that this situation exists in many other cases also, as suggested by Evenari (1949), for example in grapes. The same author reviewed the older literature regarding germination inhibitors. Since this time a number of seeds and fruits have been investigated and some inhibiting substances isolated from them. The use of chromatographic techniques similar to those used for auxins has resulted in the detection of many as yet unknown substances, but a few have been identified. Thus coumarin and hydroxy cinnamic acid and their derivatives, as well as vanillic acid, have been shown to occur in barley husks (van Sumere *et al.*, 1958). Both Varga (1957) and Ferenczy (1957) studied the inhibiting compounds occurring in lemons, strawberries and apricots. Coumarin and derivatives of cinnamic and benzoic acids were identified as the compounds responsible for inhibitory action. They suggested that the inhibitory activity is the result of the additive effect of a number of such compounds. The same type of compounds have also been identified in clusters of sugarbeet seeds and it is suggested that these substances here also act as germination inhibitors (Roubaix and Lazar, 1957; Massart, 1957). However, in the case of the inhibitory substance in sugarbeet the evidence is very uncertain and no definite conclusion can yet be drawn as to the nature of inhibitory compounds, if any, present in the seed balls.

Thus in most of the cases mentioned, it seems reasonable to accept the view that some substance or substances occur which inhibit germination. But, as already stated, the evidence regarding the chemical nature of these compounds is extremely vague, nor is it certain whether specific or non-specific compounds are involved. Evenari in his review attempts to classify the naturally occurring inhibitors found in seeds and fruits. He mentions cyanide-releasing complexes especially in Rosaceous seeds, ammonia-releasing substances, mustard oils, mainly in Cruciferae, various organic acids, unsaturated lactones, especially coumarin, parasorbic acid and protoanemonin, aldehydes, essential oils, alkaloids and phenols. From a glance at this list it can be seen that these compounds are not in any way restricted in occurrence to seeds and fruits. On the contrary substances of this general type occur in leaves, roots and other parts of plants as well as in the fruits and seeds. For this reason it is easy to understand why germination inhibitory substances have been isolated from a variety of plant tissues. Here again it is not clear what, if any, the biological importance of these substances is. This aspect of the problem will be dealt with in Chapter 7.

III. GERMINATION STIMULATORS

Various chemical substances can completely or partially substitute for light in breaking dormancy. These substances are very varied in chemical nature. Some of them are simple compounds, such as potassium nitrate and thiourea, while others are complicated molecules such as the gibberellins and kine-

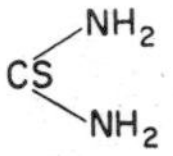

FIG. 4.5. Structure of some dormancy-breaking substances

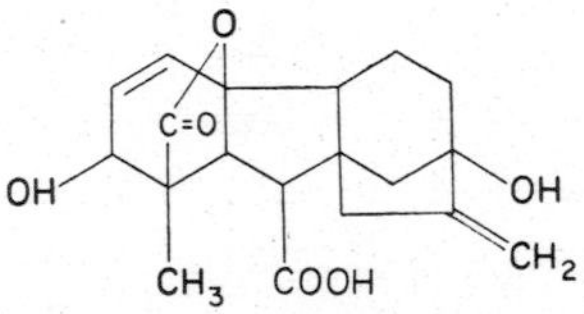

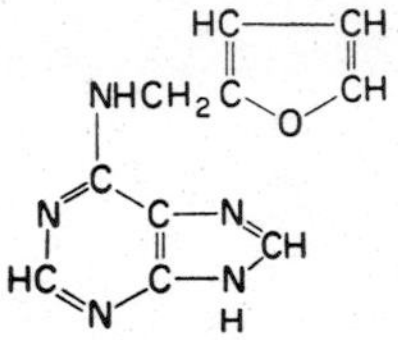

(I) Potassium nitrate; (II) Ethylene; (III) Thiourea; (IV) Gibberellin A 3; (V) Kinetin.

tin. The structure of a few of these dormancy-breaking substances is shown in Fig. 4.5.

The effect of potassium nitrate on germination was discovered when it was noted that Knop's solutions promoted the germination of some seeds. Further experiments showed that potassium nitrate was responsible for this stimulation. Potassium nitrate promotes the germination of a number of seeds in the dark, e.g. *Lepidium virginicum*, *Eragrostis curvula*,

TABLE 4.9 — EFFECT OF KNO_3 ON GERMINATION OF VARIOUS SEEDS
(After Hesse, 1924)
Concentration of KNO_3, $0 \cdot 01$ M

Seed	Temp.	Incubation time in days	Percentage germination	
			water	KNO_3
Veronica longifolia	16–20°	20	3·5	45
Hypericum perforatum	17–22°	20	28·0	57
H. hirsutum	15–20°	23	18·0	27
H. hirsutum	15–20°	28	20·0	43
Epilobium hirsutum	16–20°	14	23·0	40
E. montanum	16–20°	14	6·5	89

Polypogon monspelliensis, various species of *Agrostis* and *Sorghum halepense*. The effect on some other species is shown in Table 4.9. The stimulation obtained by potassium nitrate is dependent on its concentration as shown in Fig. 4.6. As with light, potassium nitrate stimulation shows interaction with tempera-

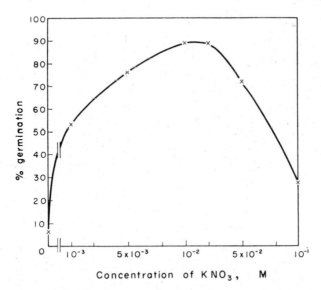

FIG. 4.6. Relation between germination percentage of seeds of *Epilobium montanum* and the potassium nitrate concentration (Germination in the dark for 14 days at 16°–20°) (After Hesse, 1924)

ture. The germination of *Eragrostis curvula* is stimulated between 15 and 30°C in the dark by 0·2 per cent potassium nitrate. At higher temperatures or alternating temperatures there was no effect. In contrast the germination of *Polypogon* is promoted by potassium nitrate only at alternating temperatures (Toole, 1938).

The dark germination of many seeds is stimulated by thiourea. However, thiourea must be applied in high concentration in order to promote germination. Thus for lettuce seeds the effective concentration is of the order of 10^{-2} to 10^{-3} M. These concentrations apply if the seeds are actually germinated in the solution of thiourea. A frequently adopted procedure is soaking the seeds in a concentrated solution of thiourea, 0·5 to 3 per cent, for a short time and then transferring them to water. Thus thiourea has been shown to stimulate the germination of seeds of *Cichorium*, *Gladiolus* (Shieri, 1941) and *Quercus*. In oak seed, as well as in other tree species such as *Larix* and *Picea*, thiourea substituted for the cold temperature treatment (Deubner, 1932; Johnson, 1946). Thiourea substitutes for the natural germination stimulator which develops in *Fraxinus* seeds during chilling, see also Fig. 4.1 (Villiers and Wareing, 1960). In some varieties of peach seeds, thiourea could substitute for after-ripening in promoting germination but subsequent seedling growth was abnormal (Tukey and Carlson, 1945). In lettuce and endive, thiourea abolished the inhibitory effect of high temperatures, while in lettuce thiourea also abolished the light requirements of the seeds. The effect of thiourea on the germination of lettuce was first described by Thompson and Kosar (1939) and has since been the subject of much study. Thiourea, in addition to stimulating germination, inhibits growth. Prolonged treatment of seeds with high concentrations of thiourea is therefore liable to cause an apparent inhibition of germination, because the emergence of the root is prevented. Thus it was found that germination of lettuce above 90% was obtained, if the seeds were kept 24 hours in 5×10^{-2} M thiourea and then transferred to water. Water controls germinated only 40 per cent. Keeping the seeds longer in thiourea reduced the germination below 90 per cent. As in many other effects on germination, here also there exist interactions between the effect of thiourea and that of other factors affecting germination, such as light and temperature. Moreover, the effects observed are a function of the thiourea concentrations (Table 4.10). It is clear that at different tempera-

TABLE 4.10—THE EFFECT OF
THIOUREA ON THE GERMINATION
OF LETTUCE, VARIETY GRAND
RAPIDS, AT DIFFERENT CONCEN-
TRATIONS AND TEMPERATURES
(Poljakoff-Mayber *et al.*, 1958)

Thiourea	%germination	
	20°C	26°C
0	26·8	5·8
$1·0 \times 10^{-2}M$	67·4	21·4
$2·5 \times 10^{-2}M$	76·5	16·1
$5·0 \times 10^{-2}M$	56·3	12·4

tures the concentration of thiourea effective in causing germination is different. An interaction between light and thiourea was shown by Evenari *et al.* (1954). The effects of thiourea on the metabolism of seeds, which have been studied in some detail, will be discussed later. The action of derivatives of thiourea on the germination of lettuce seeds has also been studied. All modifications of the structure of thiourea, e.g. ethyl, methyl and phenyl substitution, converted it from a germination stimulator to a substance inhibiting germination (Mayer, 1956). Thiourea has been shown to be present in the seeds of at least one plant species, *Laburnum anagyroides* (Klein and Farkass, 1930).

Ethylene chlorhydrin has been shown in a number of cases to have similar effects on germination as those of thiourea. However, it is usually less effective. Both substances were first used to break dormancy in buds and only at a later stage used for breaking dormancy in seeds. Ethylene is a compound

closely related to ethylene chlorhydrin, is active in breaking
dormancy in buds and is produced by many plant tissues and
especially by ripening fruit. That other compounds of stimulating activity are of widespread occurrence in nature seems
likely. Libbert and Lubke (1957) showed that the widely distributed scopoletin can promote germination in old seeds
of *Sinapis alba* at very low concentrations. Ruge (1939) indicated
that in *Helianthus annuus* seeds, there is a balance between a germination-stimulating and an inhibiting factor, without however
identifying either of them. The best established cases of natural
stimulators are those of the factors responsible for stimulating
the germination of the seeds of the parasitic plants *Striga sp.*
and *Orobanche sp.* Brown and his co-workers (1952) showed that
seeds of both these plants germinate in the vicinity of the host
only after the latter secretes a germination-stimulating substance. It appears that a number of compounds, one of which
may be xyloketose, have stimulatory activity.

IV. HORMONES IN GERMINATION

In a number of isolated cases hormones have been shown
to promote germination. Although this effect has been known
for some time, it is by no means clear-cut. The term hormone
is here used in the widest sense of the word, including various
growth substances and natural compounds having a regulatory function in the plant. Most of the dubious results have
been obtained using auxin. Much clearer effects have been
obtained by the external application of gibberellins or gibberellic acid, but the effect is by no means universal. Lona
(1956) and Kahn *et al.* (1956) were among the first to show that
gibberellic acid stimulates the germination of *Lactuca scariola*
and *Lactuca sativa* in the dark. In these cases gibberellic acid
substituted for light in promoting germination. Other cases
of substitution of gibberellic acid for red light are known in
the case of seeds of *Arabidopsis*, *Kalanchoe* and *Salsola volkensii*.
However, in a number of seeds whose germination is either

TABLE 4.11 — EFFECT OF GIBBERELLIC ACID ON THE GERMINATION OF VARIOUS SEEDS
(Compiled from data of Kallio and Piironen, 1959; Corns 1960)

Plant	Treatment with G. A.	% germination Treated	% germination Water central
Avena fatua	500 ppm	57	26
Sinapis arvensis	500 ppm	89	9
Thlapsi arvense	500 ppm	90–100	0
Gentiana nivali	1000 ppm (24 hr)	90	0
Bartschia alpina	1000 ppm (24 hr)	90	5
Draba hirta	1000 ppm (24 hr)	95	0

promoted or inhibited by light, e.g. *Juncus maritimus*, *Oryzopsis miliacea* and others, treatment with gibberellic acid was not effective in promoting germination (Leizorowitz, 1959). Recently the germination of a number of species of plants whose germination is not affected by light have been shown to be promoted by gibberellic acid, as shown in Table 4.11. The sensitivity of some of the seeds depended on the time of harvest. There was no relation between the place of the seeds in the systematic classification of plants and their response to gibberellic acid.

The effect of gibberellic acid seems to be similar to that of light in promoting germination (Table 4.12). But gibberellic acid is far more effective than red light in reversing the high

TABLE 4.12 — EFFECT OF LIGHT AND GIBBERELLIC ACID ON THE GERMINATION OF LETTUCE AT 26°C
(Evenari *et al.*, 1958)

Treatment	% germination In water	% germination In gibberellic acid $2 \cdot 9 \times 10^{-5}$ M
Dark	12·0	39·0
Red	44·0	66·5
Far-red	5·0	25·0
Red and far-red	11·0	35·0

temperature inhibition of germination in lettuce. Thus 100 ppm gibberellic acid stimulated the germination of lettuce seeds at 30°C from 2 per cent in the water to 33 per cent. In contrast red light only stimulated the germination from 2 to 4 per cent. When both gibberellic acid and light were given, the germination percentage obtained was 50. Gibberellic acid can also reverse the inhibition of germination caused by high osmotic pressure. Kahn (1960) showed that lettuce seeds germinating to 82 per cent in the dark, in water, gave only 22 per cent germination in 0·15 M mannitol. However, addition of 35 ppm gibberellic acid to the mannitol resulted in a germination percentage of 61, showing a reversal of osmotic inhibition. Similar reversals of osmotic inhibition were obtained by using red light, from 36 per cent in 0·18 M mannitol to 88 per cent in mannitol plus red light. Originally it was assumed that red light and gibberellic acid act in an entirely similar fashion in promoting germination of lettuce seeds. However, more detailed investigation along the lines mentioned, studying the interaction of gibberellic acid with far-red light and with high temperatures, showed that this was not the case (Evenari et al., 1958). Thus, it is not always possible effectively to reverse the gibberellic acid induced stimulation by the use of far-red light, indicating that part of the gibberellic acid induced germination differs from that induced by red light. These experiments have resulted in the general conclusion that gibberellic acid and red light act only partially in the same way and that their mode of action is not identical (Ikuma and Thimann, 1960; Fujii et al., 1960). That gibberellic acid might be of importance in determining the germination of seeds in nature is indicated by the isolation from Phaseolus, lettuce and many other seeds of gibberellin-like compounds (Phinney and West, 1960). There is some evidence to indicate that the amount of gibberellin-like substances changes during different stages of germination and during ripening or after-ripening of the seeds.

Another plant hormone, kinetin, also effects germination of seeds and especially of lettuce seeds. Miller (1958) showed that kinetin promotes the germination of seeds. A large number of derivatives of kinetin, where the furfuryl group is replaced by other groupings, also stimulate germination. While originally it was thought that kinetin substitutes for red light in germination, it was later shown that in fact the seeds were sensitized by kinetin so that a smaller dose of light would induce their germination. Weiss (1960) has recently shown that while in water, lettuce seeds required 3600 f.c. sec light to promote their germination, in the presence of kinetin 720 f.c. sec light were sufficient to bring about the same effect. The seeds need not be kept continuously in a solution of kinetin. As little as a few hours in the solution, at a suitable period, is sufficient to cause the sensitization. That kinetin does not substitute for light is also indicated by the fact that light-inhibited seeds such as *Oryzopsis miliacea* are stimulated by kinetin, while other light sensitive seeds such as *Amaranthus* are not effected by kinetin at all.

The effect of hormones of the indolylacetic acid, IAA, type on germination has long been in dispute. Numerous workers have investigated the effect of IAA and similar substances on the germination of a variety of seeds, and have obtained conflicting results, stimulation or inhibition being obtained, depending on the concentration of IAA and the type of seed used. However the most general effect is an absence of response of the seeds to physiological concentrations of IAA. Soeding and Wagner (1955) attempted to relate divergent observations to the state of dormancy of the seeds. However, they failed experimentally to prove such a relation for seeds of *Poa*. A relation between depth of dormancy and response to IAA was, however, established for lettuce seeds. Even here the observed absolute effects were usually very small. In seeds where germination in the dark was of the order of 4–10 per cent, IAA at a concentration of 10^{-7} M raised the germination in the dark to 20–30 per cent, while on seeds germinating 60–80 per cent in the dark, IAA had no effect whatever. In other experiments the response of the seeds to IAA was found to be temperature-dependent. At 20° 10^{-7}M

IAA raised the germination of lettuce seeds from 27 to 47 per cent, while at 26°C the germination percentage was 8, both in water and in IAA (Poljakoff-Mayber, 1958).

Thus from these results some relationship between response to IAA, dormancy and germination conditions emerges, but the effects are in no way clear-cut. It is not possible to relate these results to internal IAA concentrations, as dry lettuce seeds apparently do not contain IAA. In general, therefore, it must be concluded that IAA can, under very special conditions, stimulate germination, but normally it has little or no effect.

From the above discussion on inhibitors and stimulators it may be concluded that the germination of seeds is controlled by a variety of external and internal factors. These factors include, in addition to simple and environmental conditions, also the presence of external and internal germination-inhibiting and stimulating substances.

BIBLIOGRAPHY

AKKERMAN, A. M. and VELDSTRA, H. (1947) *Rec. Trav. Chim.* **66**, 411.

AUDUS, L. J. and QUASTEL, J. H. (1947) *Nature, Lond.* **159**, 230.

AXENTJEV, B. N. (1930) *B. B. C.* **46**, 119.

BALLARD, L. A. T. (1958) *Austr. J. Biol. Sci.* **11**, 246.

BARTON, L. V. (1934) *Contr. Boyce Thompson Inst.* **6**, 69.

BARTON, L. V. and CROCKER, W. (1948) *Twenty Years of Seed Research*, Faber & Faber, London.

BARTON, L. V. and SCHROEDER, E. M. (1942) *Contr. Boyce Thompson Inst.* **12**, 277.

BLACK, M. and WAREING, P. F. (1955) *Physiol. Plant* **8**, 200.

BROWN, R. (1940) *Ann. Bot. N. S.* **4**, 379.

BROWN, R. (1946) *Nature, Lond.* **157**, 64.

BROWN, R., JOHNSON, A. W., ROBINSON, E. and TYLER, G. (1952) *Biochem. J.* **50**, 596.

BROWN, R., GREENWOOD, A. D., JOHNSON, A. W., LANDSDOWN, A. R., LONG, A. G. and SUNDERLAND, N. (1952) *Biochem. J.* **52**, 571.

CORNS, W. G. (1960) *Canad. J. Plant Sci.* **40**, 47.

CROCKER, W. (1906) *Bot. Gaz.* **42**, 265.

CROCKER, W. and BARTON, L. V. (1953) *Physiology of Seeds*, Chronica Botanica.

DAVIS, W. E. (1930) *Amer. J. Bot.* **17**, 58.

DEUBNER, C. G. (1932) *J. Forestry* **30**, 672.

ECKERSON, S. (1913) *Bot. Gaz.* **55**, 286.

EVENARI, M. (1949) *Bot. Rev.* **15**, 143.

EVENARI, M., STEIN, G. and NEUMAN, G. (1954) *Proc. ist. Int. Photobiological Congress, Amsterdam*, p. 82.

EVENARI, M., NEUMAN, G., BLUMENTHAL-GOLDSCHMIDT, S., MAYER, A. M. and POLJAKOFF-MAYBER, A. (1958) *Bull. Res. Council Israel* **6 D**, 65.

FERENCZY, I. (1957) *Acta Biol. Hung.* **8**, 31.

FLEMION, F. (1933) *Contr. Boyce Thompson Inst.* **5**, 143.

FLEMION, F., and PROBER, P. L. (1960) *Contr. Boyce Thompson Inst.* **20**, 409.

FLEMION, F. and DE SILVA, D. S. (1960) *Contr. Boyce Thompson Inst.* **20**, 365.

FUJII, T., ISIKAWA, S., and NAKAGAWA, A. (1960) *Bot. Mag. Tokyo* **73**, 404.

GRANT LIPP, A. E. and BALLARD, L. A. T. (1959) *Austr. J. Agr. Res.* **10**, 495.

HAMLY, D. H. (1932) *Bot. Gaz.* **93**, 345.

HESSE, O. (1924) *Bot. Archiv.* **5**, 133.

IKUMA, H. and THIMANN K. V. (1960) *Plant Phys.* **35**, 557.

ISIKAWA, S. (1954) *Bot. Mag. Tokyo* **67**, 51.

ISIKAWA, S. (1955) *Kumamoto J. Science Ser.* **B 2**, 97.

ISIKAWA, S. and ISHIKAWA, T. (1960) *Plant and Cell Physiol.* **1**, 143.

JOHNSON, L. P. V. (1946) *Forestry Chronicle* **22**, 17.

KAHN, A., GOSS, J. A. and SMITH, D. E. (1956) *Plant Phys.* **31**, suppl. 37.

KAHN, A. (1958) *Plant Phys.* **33**, 115.

KAHN, A. (1960) *Plant Phys.* **35**, 1.

KAHN, A. (1960) *Plant Phys.* **35**, 333.

KALLIO, P. and PIIRONEN, P. (1959) *Nature, Lond.* **183**, 1930.

KHUDAIRI, A. K. (1956) *Physiol. Plant* **9**, 452.

KIDD, F. (1914) *Proc. Roy. Soc.* **B 87**, 609.

KLEIN, G. and FARKASS, E. (1930) *Öst. Bot. Z.* **79**, 107.

KOLLER, D. (1954) Ph. D. Thesis Hebrew University, Jerusalem (in Hebrew).

LEGGATT, C. W. (1946) *Can. J. Research* **C 24**, 7.

LEIZOROWITZ, R. (1959) M. Sc. Thesis, Jerusalem (in Hebrew).

LERNER, H. R., MAYER, A. M. and EVENARI, M. (1959) *Physiol. Plant* **12**, 245.

LIBBERT, E. and LUBKE, H. (1957) *Flora* **145**, 256.

LIBBERT, E. (1957) *Phyton* **9**, 81.

LONA, F. (1956) *Ateneo Parmense* **27**, 641.

MAIER, W. (1933) *Jahr. Wiss. Bot.* **78**, 1.

MANCINELLI, A. (1958) *Ann. di Botanica* **26**, 56.

MASSART, L. (1957) *Biochimia (USSR)* **22**, 117.

MATO, N. (1924) *Sitz. Acad. Wiss. Matt. Natur. K. L.* **133**, 625.

MAYER, A. M. and EVENARI, M. (1952) *J. Exp. Bot.* **3**, 246.

MAYER, A. M. and EVENARI, M. (1953) *J. Exp. Bot.* **4**, 257.

MAYER, A. M. (1956) *J. Exp. Bot.* **7**, 93.

MILLER, C. O. (1958) *Plant. Phys.* **33**, 115.

MISRA, G. and PATNAIK, S. N. (1959) *Nature, Lond.* **183,** 989.

NAGAO, M., ESASHI, Y., TANAKA, T., KUMAIGAI, T. and FUKUMOTO, S. (1959) *Plant and Cell Phys.* **1,** 39.

NIETHAMMER, A. (1927) *Biochem. Z.* **185,** 205.

NUTILE, G. E. (1945) *Plant Phys.* **20,** 433.

OLNEY, H. O. and POLLOCK, B. M. (1960) *Plant Phys.* **35,** 970.

OPPENHEIMER, H. (1922) *Sitz. Akad. Wiss. Wien Abt.* **1 131,** 279.

PHINNEY, B. O. and WEST, C. A. (1960) *Ann. Rev. Plant Phys.* **11,** 411.

POLJAKOFF-MAYBER, A. (1958) *Bull. Res. Council Israel.* **6D,** 78.

POLJAKOFF-MAYBER, A., MAYER, A. M. and ZACHS, S. (1958) *Ann. Bot. N. S.* **22,** 175.

POLLOCK, B. M. and OLNEY, H. O. (1959) *Plant. Phys* **34,** 131.

ROUBAIX, J. DE and LAZAR, O. (1957) *La Sucrerie Belge* **5,** 185.

RUGE, U. (1939) *Z. f. Bot.* **33,** 529.

SHIERI, H. B. (1941) *The Gladiolus* **16,** 100.

SHULL, C. A. (1911) *Bot. Gaz.* **52,** 455.

SHULL, C. A. (1914) *Bot. Gaz.* **57,** 64.

SOEDING, H. and WAGNER, M. (1955) *Planta* **45,** 557.

SPAETH, J. N. (1932) *Amer. J. Bot.* **19,** 835.

STEARNS, F. and OLSEN J. (1958) *Amer. J. Bot.* **45,** 53

SUMERE, C. F. VAN, HILDERSON, H. and MASSART, L. (1958) *Naturwiss.* **45,** 292.

SUMERE, C. F. VAN (1960) In *Phenolics in Plants in Health and Disease,* p. 25, Pergamon Press, Oxford.

THOMPSON, R. C. and KOSAR, W. F. (1939) *Plant Phys.* **14,** 561.

THORNTON, N. C. (1935) *Contr. Boyce Thompson Inst.* **7,** 477.

TOOLE, E. H. (1959) In *Photoperiodism and related phenomena in plants and animals,* AAAS Publ. No. 55, (ed. Withrow).

TOOLE, E. H., TOOLE, V. K., BORTHWICK, H. A. and HENDRICKS, S. B. (1955) *Plant Phys.* **30,** 473.

TOOLE, V. K. (1938) *Proc. Ass. Off. Seed Analyst.* N. America. 30th Ann. Meeting. p. 227.

TUKEY, H. B. and CARLSON, R. F. (1945) *Plant Phys.* **20,** 505.

UHVITS, R. (1946) *Amer. J. Bot.* **33,** 278.

VARGA, M. (1957) *Act. Biol. Hung.* **7,** 39.

VARGA, M. (1957) *Acta Biol. Szeged* **3,** 213.

VARGA, M. (1957) *Acta Biol. Szeged* **3,** 225.

VARGA, M. (1957) *Acta Biol. Szeged* **3,** 233.

VILLIERS, T. A. and WAREING, P. F. (1960) *Nature, Lond.* **185,** 112.

WAREING, P. F. and FODA, H. A. (1957) *Physiol. Plant.* **10,** 266.

WEISS, J. (1960) *C. R. Acad. Sci. Paris* **251,** 125.

WIESNER, J. (1897) *Ber. Deut. Bot. Ges.* **15,** 503.

WYCHERLEY, P. R. (1960) *J. Rubb. Res. Inst. Malaya* **16,** 99.

CHAPTER 5

METABOLISM OF GERMINATING SEEDS

THE dry seed is characterized by a remarkably low rate of metabolism. This is probably a direct result of the almost complete absence of water in the seed, whose water content is of the order of 5–10 per cent. Despite this almost complete absence of metabolism in the seed, it cannot be assumed that it lacks the potentiality for metabolism. When dry seeds are broken up by grinding in a suitable aqueous medium it is possible to show in the extract the presence of a considerable number of enzyme systems. Thus it can be concluded that the dry seed is a well equipped functional unit which can carry out a large number of biochemical reactions, provided that the initial hydration of the proteins and more specifically of the enzyme proteins, has taken place.

One of the most frequently used criteria for determining the rate of metabolism is the rate of respiration. Dry seeds show a very low rate of respiration, which begins to rise rapidly when the seeds are placed in water. Thus the first easily observable metabolic change in the seed, well before germination occurs, is the increase in the respiratory rate from values close to zero to appreciable rates. The second effect, which is very easily observed, is the breakdown of reserve materials in the seed. This reaction again is initiated by the hydration of the proteins. As long as the seeds are dry almost no changes whatever are observed in their chemical composition. As soon as the seed is hydrated very marked changes in composition in its various parts occur. These changes occur even when the seed is placed in water only, without any nutrients, and

101

in complete absence of assimilation. The chemical changes which occur are complex in nature. They consist of three main types: the breakdown of certain materials in the seed, the transport of materials from one part of the seed to another and especially from the endosperm to the embryo, or from the cotyledons to the growing parts, and lastly the synthesis of new materials from the breakdown products formed. The only substances normally taken up by seeds during germination are water and oxygen. In many instances substances are lost from the seed during initial stages of germination. The initial stages of germination are consequently accompanied by a net loss of dry weight due to the oxidation of substances on the one hand and leakage out of the seed on the other hand. Only when the seedling has formed, i.e. when a root is present which takes up minerals, and the cotyledons or first leaves are exposed to light and capable of photosynthesis, does an increase in dry weight begin. In the following, we will discuss the breakdown of reserve materials in seeds, their respiration and the general metabolism of the various seed components. The effect of germination stimulators and inhibitors on metabolism will be discussed in a separate chapter.

I. Changes in Storage Products during Germination

As already mentioned in Chapter 2, the chief types of seeds are those containing fat and those containing carbohydrates as a reserve material. In both cases varying amounts of proteins are also present. Part of the proteins constitutes the various enzyme systems and the remainder is present as a reserve material. A variety of other materials are also present in the dry seeds as already discussed in Chapter 2. Although some of these compounds are present in small amounts only, their metabolism is nevertheless of considerable importance during germination. A number of other compounds do not seem to be metabolized to any appreciable extent.

A very clear illustration of the changes in composition in different parts of a seed during germination is given in Figs. 5.1–5.8. These figures are taken from the detailed investigations of Oota *et al.* (1953) on the metabolism of germinating beans, *Vigna sesquipedalis*. This seed is characterized by large

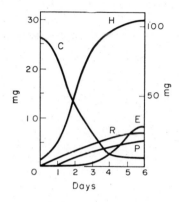

FIG. 5.1. Changes in dry weight of embryonic organs of *Vigna sesquipedalis* during germination. Values per individual are plotted. P – a pair of plumules; E – epicotyl; H – hypocotyl; R – radical; C – a pair of cotyledons.

Scales on the right-hand ordinate for C; on the left hand ordinate for P, E, H and R. (From Oota *et al.*, 1953)

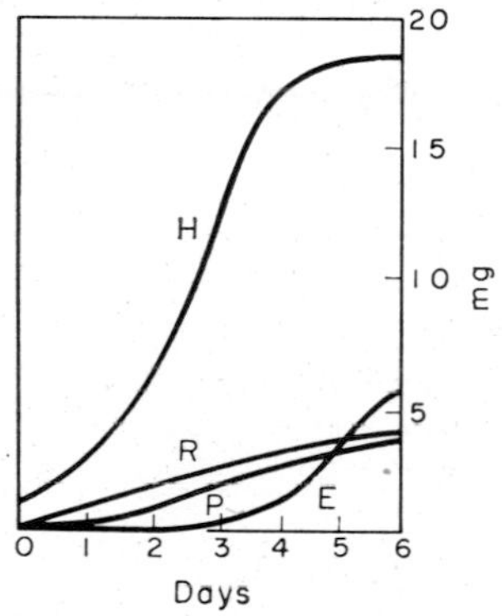

FIG. 5.2. Changes in content of water-soluble matter of embryonic organs of *Vigna sesquipedalis* during germination. Values per individual are plotted. Symbols as Fig. 5.1. (From Oota *et al.*, 1953)

cotyledons and epigeal germination. The cotyledons show a loss of all compounds studied, while all the other organs show an increase in the various seed constituents. Particularly striking is the increase of materials in the rapidly growing hypocotyl, which ceases as the hypocotyl ceases to grow. Very few such detailed studies have been made and the work of Oota is therefore particularly interesting. The seed studied by Oota lends itself particularly to such detailed analysis as it is techni-

cally feasible to differentiate between hypocotyl, root, cotyledons, plumule and epicotyl.

As will be seen from Fig. 5.1 the cotyledons show a
steady decrease in dry weight for the first four days of germination. At the same time there is a similar increase in dry
weight in the other parts of the seedling and especially the

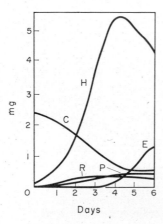

FIG. 5.3. Changes in content of
soluble sugar, expressed as
equivalents of glucose, of embryonic organs of *Vigna sesquipedalis* during germination. Values per individual are plotted.
Symbols as in Fig. 5.1. (From
Oota *et al.*, 1953)

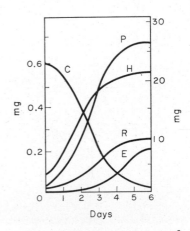

FIG. 5.4. Changes in content of
insoluble polysaccharide (expressed as equivalents of glucose) of embryonic organs of
Vigna sesquipedalis during germination. Values per individual
are plotted. Symbols as in
Fig. 5.1. Scales on the right
hand ordinate for C; on the
left hand ordinate for P, E, H,
and R. (From Oota *et al.*, 1953)

hypocotyl. When the latter ceases growing, after three or four
days, a rapid rise in dry weight of the epicotyl is observed. All
marked changes in the epicotyl are delayed till the onset of
its growth. During the six days of measurement about 20 per
cent dry weight is lost from the seed, presumably due to respiration. A study of Fig. 5.2 and Fig. 5.8 shows a very similar situation for total water soluble and total water insoluble
material in the embryonic organs. Carbohydrates accumulate

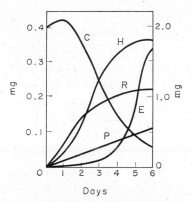

FIG. 5.5. Changes in content of soluble nitrogen of embryonic organs of *Vigna sesquipedalis* during germination. Values per individual are plotted. Symbols as Fig. 5.1. Scales on right hand ordinate for H; on left hand ordinate for P, E, R and C. (From Oota *et al.*, 1953)

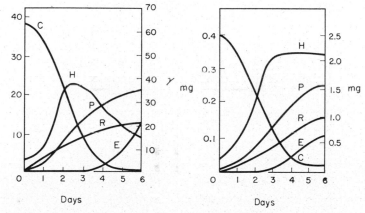

FIG. 5.6. Changes in content of pentose nucleic acid phosphorus of embryonic organs of *Vigna sesquipedalis* during germination. Values per individual are plotted. Symbols as in Fig. 5.1. Scales on the right hand ordinate for C; on the left hand ordinate for P, E, H and R. (From Oota *et al.*, 1953)

FIG. 5.7. Changes in content of protein nitrogen of embryonic organs of *Vigna sesquipedalis* during germination. Values for individual are plotted. Symbols as in Fig. 5.1. Scales on the right hand ordinate for C; on the left hand ordinate for P, E, H and R. (From Oota *et al.*, 1953)

in the various parts of the embryo in a similar fashion, but as will be seen in Fig. 5.3, after four days soluble sugars again move out from the hypocotyl and begin to appear in the epicotyl. Insoluble polysaccharides, consisting primarily of starch and dextrin, however, do not show this decrease in the hypocotyl (Fig. 5.4). The pattern for soluble and protein nitrogen (Fig. 5.5 and Fig. 5.7) is again very similar, but the increase in protein nitrogen lags well behind the increase in soluble nitrogen. The decrease in soluble nitrogen of the cotyledons does not begin till the second day. In fact, soluble nitrogen shows a distinct peak after two days, presumably due to an unequal rate of protein breakdown and transport of the breakdown products from the cotyledons. Figure 5.6 shows the changes in RNA (ribose nucleic acid) in the different parts of the embryo. The cotyledons lose steadily but the hypocotyl shows a peak in RNA between the second and third day and a subsequent linear decrease. The peak appears before the cessation of hypocotyl growth. The data brought above show a marked mobility of the various components of the seed. The function of the cotyledons as storage organs which empty out as other parts of the embryo develop, is clearly demonstrated.

There are few other data as complete as those of Oota *et al.* on the behaviour of the various chemical constituents of seed during germination. The behaviour of the seeds having an endosperm as the main storage organ have been examined to some extent. Here the general trend is of a similar nature, showing a movement of material from the endosperm and a corresponding increase in the embryo.

In rice, for example, Fukui and Nikuni (1956) showed that there was a total loss of dry weight of 20–25 per cent. The endosperm showed an overall decrease in dry weight and the root and shoot a corresponding increase. The increase in dry weight of the root lagged behind that of the shoot, as in rice, the coleoptile grows out before the roots. Starch showed a marked decrease in the endosperm, but neither root nor shoot showed a corresponding accumulation of starch. The products resulting from the breakdown of starch were not examined.

A similar survey was made by Yocum (1929) for wheat seeds. He grew wheat in soil and analysed the composition of both the entire seed or seedling and, after separation into different organs was possible, also the composition of the endosperm, root and plumule. Table 5.1 shows some of the results obtained. It will be noted that starch decreased continuously in the endosperm as well as in the whole plant. Dextrins also

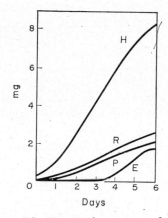

Fig. 5.8. Changes in content of water insoluble matter of embryonic organs of *Vigna sesquipedalis* during germination. Values per individual are plotted. Symbols as in Fig. 5.1. (From Oota *et al.*, 1953)

disappeared from the endosperm, although on the first day some dextrins are formed from other substances. Fats also decreased during the first few days, but later fats are reformed. The resynthesis of fats and the appearance of reducing and other sugars is presumably the result of transformations of the disappearing starch. Nitrogen also leaves the endosperm and appears in other parts of the seedling.

Oil-containing seeds have been investigated by Yamada (1955). He showed the disappearance of lipids in the various parts of the seeds of the castor bean during germination. Total lipids decreased both in the endosperm and the cotyledons, as

TABLE 5.1 — CHANGES IN COMPOSITION OF WHEAT SEEDS DURING GERMINATION AND GROWTH

The table illustrates the change in storage materials during germination and growth. The data show both the composition of the entire seed or seedling and also the composition of the endosperm, root and plumule after different times of germination. The results are given as weight of substances in mg per 100 seeds or parts of seedlings.

(Compiled from Yocum, 1925)

Time in days	Plant Part	Dry weight	Ether extract (lipids)	Reducing sugars	Total sugar	Dextrin	Starch	N_2
0	Original Seed	2685	66.9	0	53.9	43.5	1781.0	59.1
1	Seed	2708	63.1	0	44.7	82.0	1621.3	54.4
2	Seed	2593	57.6	22.8	137.2	111.0	1079.0	46.2
3	Seed	2544	51.0	83.1	121.4	81.3	1343.5	48.8
6	Seedling	2476	45.9	234.8	465.8	120.5	472.1	49.6
9	Seedling	2383	95.3	117.3	208.9	86.5	117.5	63.3
12	Seedling	2031	94.7	15.6	37.1	19.9	20.2	54.3
3	Plumule	110	2.1	3.1	9.2	3.3	2.6	5.9
6	Plumule	383	8.7	13.6	20.4	9.6	2.9	14.8
9	Plumule	875	54.8	13.6	26.5	11.7	7.0	37.7
12	Plumule	1054	68.3	0	4.1	3.0	0	34.8
3	Root	103	2.6	3.6	6.3	2.5	1.5	5.4
6	Root	286	6.1	8.5	14.8	5.2	0.6	7.5
9	Root	469	10.7	11.7	15.5	6.8	1.5	12.9
12	Root	691	10.2	7.2	12.5	7.8	2.3	15.9
3	Remaining Seed (endosperm)	2332	46.3	76.5	105.9	75.4	1339.5	37.6
6	(endosperm)	1807	31.2	212.7	430.7	105.7	468.0	27.3
9	(endosperm)	839	29.7	102.1	167.0	60.1	109.0	12.8
12	(endosperm)	287	16.3	8.4	20.5	8.9	17.9	3.6

did the total amount of lipids as germination proceeds (Fig. 5.9). In contrast carbohydrates began to appear in the endosperm as the fats disappear, showing accumulation up to 4 days. After this, carbohydrates disappeared again from the endosperm and appeared in the hypocotyl. This is true both for

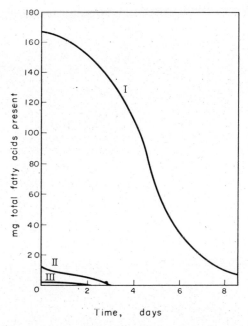

FIG. 5.9. Disappearance of fatty acids in various parts of the seed during germination of castor bean seed (Yamada, 1955). I – Endosperm; II – Cotyledons; III – Embryo.

reducing and non-reducing sugars (Fig. 5.10). It appears that fats are converted to sugar in the endosperm itself, and later the carbohydrates are transferred to the embryo.

Semenko (1957) investigated the changes in nucleic acids occurring during germination of wheat and oat seeds. The amount of RNA and DNA in the endosperm decreased and it increased in the embryo and subsequently in the seedling. The total amount of nucleic acids accumulating in the seedling after 10 days is greater than the initial amount present in the endosperm, despite the fact that the endosperm still contained some nucleic acids. Semenko concludes that nucleic acids are not only transported from the endosperm to the seedling or embryo but are also synthesized *de novo* in the seedling (Table 5.2).

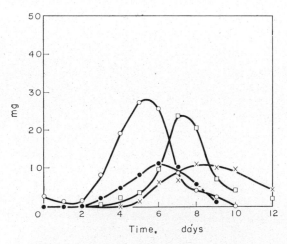

FIG. 5.10. Changes in reducing and non-reducing sugars during the germination of castor beans (Yamada, 1955).

O——O Non-reducing sugar in endosperm; ●——● Reducing sugars in endosperm;

□——□ Non-reducing sugar in hypocotyl; ✕——✕ Reducing sugars in hypocotyl.

An interesting approach to the study of mobility of seed components is that of McConnell (1957). McConnell obtained radioactive wheat seeds by feeding the parent plant with acetate, labelled in carbon 1 or carbon 2 with C^{14} He then fractionated the seeds and determined the specific activity of various seed fractions. After germinating the seeds for 5–7 days, he again determined the specific activities of the various fractions in the kernel, roots and stem of the seedling. He was able to show that a considerable part of the carbon, 17 per cent, was lost as carbon dioxide during respiration. In the remaining C^{14} he found a redistribution of labelled carbon, showing increases in specific activity in some fractions and decreases

TABLE 5.2 — CHANGES IN NUCLEIC ACID CONTENT IN
ENDOSPERM AND EMBRYO OR SEEDLING OF GERMI-
NATING WHEAT AND OATS

The results are given as mg pyrophosphate per 100
endosperms or seedlings (after Semenko, 1957).

| | | Nucleic acid content | | | |
| | | Endosperm | | Seedling | |
Time in days		RNA	DNA	RNA	DNA
Wheat	0	3·28	1·85	2·15	1·02
	2	2·89	1·64	2·76	1·22
	6	1·28	1·57	5·76	2·03
	10	1·19	0·98	7·17	4·85
Oats	0	2·14	0·45	0·97	0·59
	6	0·44	0·62	1·15	1·38
	10	0·18	0·91	3·53	1·86

in others, again indicating the mobility of the constituents of
the seed and their movement from the seed kernel to the stem
and roots, which he analysed separately. He was also able to
show that part of the protein was being respired, apparently via
glutamic acid.

In most seeds only the overall changes in certain components have been studied. Most of these clearly indicate
which materials in the seeds show a loss or gain.

II. BREAKDOWN AND METABOLISM OF
STORAGE PRODUCTS AND ENZYME ACTIVITY

From the discussion on the changes in storage products
it can be concluded that some of them undergo very marked
metabolic changes. Not necessarily those substances which are
present in the greatest amounts are broken down first. The
metabolic changes occurring in the early stages of germination
are the result of the activity of various enzymes. The enzymes
which are hydrolytic or transfer enzymes are either present in

the dry seed or very rapidly become active as the seed imbibes water. In view of the importance of some of these enzymes the reactions which are catalysed by them will be briefly discussed.

Generally, enzymes breaking down starch, proteins, hemicelluloses, polyphosphates, lipids and other storage materials, rise in activity fairly rapidly as germination proceeds, although there is no reason to think that these changes are the direct cause of the actual germination process. Changes in the activity of various enzymes have been studied in many seeds but the same general trend is observed.

1. *Carbohydrates*

Starch is normally broken down by amylases. However the exact path by which these amylases function is different in different seeds. Usually seeds contain both amylose and amylopectin. Dry seeds contain primarily β-amylase which detaches units of maltose from the starch molecule leaving dextrin, in the case of attack on amylopectin, and breaking down the molecule more or less completely in the case of amylose. In most seeds, as germination proceeds, other enzymes attacking starch are liberated. More particularly α-amylase is formed, but in addition debranching enzymes also appear. Small amounts of α-amylase are present in dry kernels of wheat. Phosphorylases also appear to be present, both in dry seeds and in germinated ones. Obviously the precise course of starch breakdown in any given seed will be determined by the relative amounts of these enzymes present, as well as by the ratio of the two main forms of starch, amylopectin being broken down preferentially. The result of the hydrolysis of starch by β-amylase, maltose, rarely accumulates as such in the seeds. Maltose is further broken down to glucose by maltase. Attack on starch by α-amylase results in a mixture of sugars, including maltose and glucose. The maltose again is usually broken down further.

The gradual disappearance of starch from rice, wheat and bean seeds has already been mentioned. For rice it was shown experimentally that the first starch component to be attacked was the amylopectin. *Zea mays* seeds contain only β-amylase in the dry seed. The rise in amylase activity in the seed during germination is primarily in α-amylase which, when amylolytic activity is at its peak, accounts for 90 per cent of total activity of the endosperm. The α-amylase originates in the scutellum

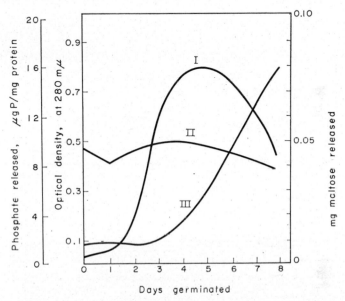

FIG. 5.11. Pattern of enzymatic activity in the cotyledons of germin-ated pea seeds (Young and Varner, 1959). I – Phosphatases; II – Pro-tease; III – Amylases.

and is secreted into the endosperm, while β-amylase appears to form only in the endosperm (Dure, 1960). In barley seeds it was found that the rise in β-amylase activity of the endosperm depended on the presence of the embryo. Removal of the latter resulted in a drop in amylase activity (Kirsop and Pollock, 1957). The rapid rise in the amylase activity as ger-mination proceeds is illustrated in Fig. 5.11. It will be seen that amylase activity begins to rise after three days, at which stage the seedling has already a root 2·5 cm in length. Young and Varner (1959) concluded that the rise in enzyme activity was connected with net protein synthesis. No clear connexion between actual germination and rise in enzyme activity was evident, as germination preceded the rise in enzyme activity. Low initial amylase activity was present already in the dry seed.

The changes in the carbohydrates of barley during germination have been studied in detail because of their importance in the malting process. Glucose and fructose rise very considerably up to six days germination at 21°C and then begin to fall again. After six days the seedlings were 5–7 cm long. Other sugars also showed marked changes. Thus maltose rose from about 1 mg per gram dry seeds to more than 55 mg after seven days of germination. Almost as great an increase occurred in oligosaccharides containing more than three hexose residues. Sucrose showed far smaller and less regular increases, as did glucodifructose. The total trisaccharides, i.e. raffinose and maltotriose, stayed more or less steady for the first five days of germination and then rose steeply, increasing five-fold in the next two days. Apparently sucrose, raffinose, glucodifructose and fructosans are associated with respiration, while the other sugars are the result of starch breakdown (McLeod, Travis and Wreay, 1953). Raffinose metabolism in barley was studied in greater detail by McLeod (1957). Raffinose was absent from the endosperm of the barley studied but accounted for 9 per cent of the dry weight of the embryo. It was rapidly utilized by the embryo under normal conditions but in isolated embryos this metabolism was retarded. Sucrose increased in the same period. No changes in either sucrose or raffinose occurred during the first twenty-four hours of germination if the seeds were kept immersed in water, i.e. steeped as in the malting process. Their metabolism seems to be closely connected to aerobic processes.

Since the endosperm of the cereal seeds is connected to the seedling by the scutellum, the latter is of some interest in considering the metabolism of such seeds during germination. Edelman *et al.* (1959) studied the function of the scutellum in considerable detail. They were able to show that glucose is removed from the endosperm, converted to sucrose in the scutellum and transported as such to the embryo. The scutellum always has a low hexose and high sucrose content, the reverse being true for the endosperm and for the seedling. Even the isolated scutellum can readily form sucrose from hexose. Sucrose was formed by a complex mechanism. Glucose is phosphorylated in the six position in the presence of ATP. Part of the glucose-6-phosphate formed is converted to fructose-6-phosphate, F-6-P, and part to glucose-1-phosphate. The glucose-1-phosphate is converted to uridine diphospho-glucose, UDPG, in the presence of uridine triphosphate, UTP. Sucrose is then formed by the condensation of UDPG and F-6-P. The enzymes necessary for all these reactions could also be demonstrated in the scutellum.

Other enzyme systems concerned with carbohydrate metabolism were shown by Lechevallier (1960) in *Phaseolus vulgaris* seeds. In these, α-galactosidase is present in appreciable amounts in the embryo and of low activity in the cotyledons of the dry seeds. During germination it falls in activity in the embryonic axis and rises in the cotyledons. An interesting feature of metabolism in *Phaseolus* seeds is the formation of malonic acid during germination. In the dry seeds this is absent or present in trace amounts, while after 5 days germination appreciable quantities are found in the embryonic axis (Duperon, 1960). In the same seeds other organic acids were also metabolized. Citric acid decreases during germination while malic acid accumulates, marked changes again occurring after 5 days. However, in *Zea mays*, citric, malic and aconitic acid all increase, though at different rates, during germination. Other tricarboxylic acid cycle intermediaries were found, if at all, only in very small amounts in a number of seeds (Duperon, 1958).

2. *Lipids*

Fats and oils are broken down in the first instance by the action of lipases. Lipases are rather non-specific esterases which cleave the bond between the fatty acids and the glycerol which esterifies them. Normally neither of the breakdown products of hydrolysis of lipids accumulates in the seeds. The fate of the glycerol which is formed is not known with certainty. It seems however to become part of the general carbohydrate pool present in the seed and as such becomes available for various processes including respiration. Thus enzyme systems have been shown in *Arachis* cotyledons which convert glycerol to glycerol phosphate which is then converted to triose phosphate. This can then be either converted to pyruvic acid or to sugars (Stumpf and Bradbeer, 1959). The fatty acids formed following lipase action accumulate in small amounts. The bulk of the fatty acids is, however, broken down further by one of a number of reactions. The fatty acids may be broken down by the process of β-oxidation, resulting in the cleavage of two carbon units in the form of acetyl, which can enter the tricarboxylic acid cycle. This reaction requires both CoA and ATP. The fatty acids may also be broken down by α-oxidation. In this process the fatty acid is peroxidatively decarboxylated and carbon dioxide formed. The long chain aldehyde is oxidized to the corresponding acid by a reaction linked to DPN. In many seeds disappearance of fats is accompanied by the appearance of carbohydrates. This reaction apparently proceeds as follows. The fatty acids undergo β-oxidation. The acetyl CoA formed is converted to malate via the glyoxylate cycle. The malate thus formed is converted to carbohydrate by a number of reactions. All these reactions have been shown to occur in the cotyledons or endosperm of fat-containing seeds such as soybeans, castor beans and groundnuts *(Arachis)*. Another enzyme which is believed to play a part in fatty acid oxidation is lipoxidase. This enzyme, which also occurs in seeds, is supposed to break the fatty acid chain into two smaller parts by peroxidative attack at a double bond. The precise function of this system is at present in doubt.

The fat content of seeds changes during germination as already mentioned. The detailed course of fat metabolism has been followed by Hardman and Crombie (1958) and Boatman and Crombie (1958) in two different seeds, *Citrullus vulgaris* and *Elaeis guineensis*. In *Citrullus* seeds there is a rapid break-down of lipids both in the cotyledons and in the rest of the seed. These lipids are largely used in respiration and do not seem to be converted to carbohydrates. Certain differences in the course of lipid metabolism of seeds grown in the dark and in the light were found. Thus in the light all the fats were broken down at an equal rate. In the dark however there seems to be a greater disappearance of linolenic acid than in the light. Another difference in fat metabolism in the light and the dark was the rate of appearance of phosphatides. In the light these formed continuously for fifteen days, while in the dark they formed for the first seven or eight days and then dropped again. However, photosynthesis occurs in these seed-lings fairly quickly. The difference observed may therefore be due to processes of assimilation rather than an essential dif-ference in behaviour of germination in light and dark.

A rather different situation was found for *Elaeis* seeds. In these seeds the lipids are located in the endosperm, which is invaded by an haustorium during germination. Free fatty acids accumulate in the endosperm but not in the haustorium. Although lipids are found in the haustorium, they appear in the esterified form. It seems that free fatty acids are transfer-red from the endosperm to the haustorium and are immedi-ately re-esterified there. In *Elaeis* seeds the bulk of the lipids is lost during germination, apparently in respiration, and little or no conversion to carbohydrate occurs. All fatty acids are metabolized at about the same rate, but saturated fatty acids disappear a little more rapidly than unsaturated ones.

Lipase has been frequently claimed to be absent from oil-containing seeds. These results can probably be ascribed to the insolubility of lipases. In order to show the presence of lipase activity it is usually necessary to defat the seeds and then activate the enzyme. Such activation is done by treating the defatted powder with dilute acetic acid or with a solution of calcium chloride. This activation may liberate the enzyme from an inactive form. The usual way of assaying lipase activity is to use the defatted powder of the seeds as an enzyme source

and an emulsion of a suitable oil as a substrate. As already
stated, lipases are rather non-specific in their action. As an
example, a crude lipase prepared from *Brassica campestris* (rape
seed) was able to hydrolyse tributyrin, triacetin and triolein as
well as natural olive, sunflower and rape oils. However, triolein was attacked more slowly than any of the other sub-

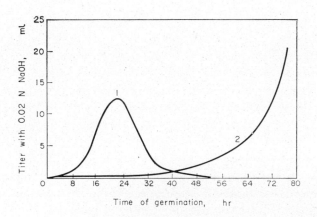

FIG. 5.12. Lipase activity in castor bean seeds during germination
(Yamada, 1957).
1 – Neutral lipase in the embryo
2 – Neutral lipase in the endosperm

strates. Triacetin was attacked at about half the rate of tributyrin (Wetter, 1957). The author points out that some sort of
specificity in the hydrolytic attack on the different oils seems
to be involved and that the observed differences were not
merely caused by differences in the solubility or degree of
dispersion of the oils.

The lipase activity of seeds changes with germination time.
It is frequently found that more than one lipase is active. These
are differentiated in various ways including the pH at which
they function. In this way it is possible to distinguish between
neutral and acid lipases. Figure 5.12 shows the changes in activity in the neutral lipase in the endosperm and the embryo
of germinating castor beans *(Ricinus communis)*. Yamada (1957)

investigated the fat metabolism of these seeds. He found, in addition to the neutral lipase, an acid lipase which is already present in the endosperm of the dry seed.

3 Proteins

As already stated, all seeds contain a certain amount of protein. In Figs. 5.5 and 5.7 were shown the breakdown of protein in cotyledons of beans on the one hand, and the appearance of new proteins in other parts of the seedling on the other hand. Other nitrogenous compounds also appear as germination proceeds. Relatively little is known as to the exact nature of the mechanism by which proteins in the seeds are broken down. It may, however, be assumed that proteins are broken down due to attack by proteases. Proteases are enzymes attacking proteins by cleavage of the peptide bond in protein molecule. The differentiation between proteinases and polypeptidases which attack smaller sub-units of the protein is not simple. These enzymes do not simply differ in the size of the molecule attacked. Rather they differ in the specificity of their attack on certain peptide linkages. Specificity is determined by such factors as the adjacent end groups, amino acid side chains near the linkage, and the relation of the size of the molecule to the number of free endgroups. For this reason and since certain proteinases can also attack peptides, any distinction between such enzymes is today artificial. In any case there is as yet very little concrete evidence as to the functioning of proteinases and peptidases in germinating seeds. The occurrence of enzymes capable of hydrolysing proteins such as gelatin has been shown both in dry seeds and in seedlings. An increase in the activity seems to occur as germination proceeds. Although the mechanism of protein breakdown is not clear, there is much evidence relating to the fate of the breakdown products. Usually there is little change in the total nitrogen content of the seed or seedling during germination, although slight losses may occur, especially due to leaching out of nitrogenous substances. Nitrogen appears to be very carefully conserved. In place of the protein broken down there appear free amino acids and amides. The mechanism by which these substances are formed, especially if seeds are germinated in the dark in water, has been the subject of extensive research. In most cases only the breakdown of proteins and the simultaneous appearance of amides has been followed. Table 5.3 shows some of these changes in *Phaseolus mungo* seeds.

TABLE 5.3 — CHANGES IN PROTEIN CONTENT AND OTHER NITROGENOUS COMPOUNDS IN *Phaseolus Mungo* SEEDS
(From Damodaran *et al.*, 1946)
The results are given as percentage of total nitrogen

Age of seedlings in days	1	4	7	10	13	16
Protein N (Extractable)	83·0	35·0	20·0	9·0	13·0	7·0
Ammonia N	0·3	0·9	1·2	0·8	1·1	2·5
Asparagine amide N	0·7	9·3	11·3	12·7	12·5	11·7
Glutamine amide N	0·3	1·1	1·4	2·3	2·9	2·5
Amine N	3·2	15·4	23·7	23·7	22·5	20·8
mg Total N/300 seedlings	534·5	486·2	511·2	517·2	499·6	505·5

Tazakawa and Hirokawa (1956) showed changes in protease activity in soybeans during germination. Activity was very low in the seeds. It rose in the cotyledons for about six days and then dropped again. The shoot began to show protease activity after approximately six days (Fig. 5.13) when it was already about 20 cm long. The presence of enzymes hydrolysing gelatin does not imply that such enzymes are of significance in the plant. However, Tazakawa *et al.* showed that the enzyme from soybeans did in fact hydrolyse proteins prepared from the same beans.

The early work on nitrogen metabolism both by Paech (1935) and Prianishnikov (1951) showed that it is possible to induce amide formation in germinating seeds in the dark by feeding them with ammonium salts. This had some sparing action on protein breakdown. If sugars were fed to the seedlings this led to a marked sparing action on protein breakdown. Simultaneous feeding of both glucose and ammonia nitrogen led to very greatly increased amide formation. This work, which has been reviewed in detail by Chibnall (1939), has led to the following view on amide formation. During germination in the dark, proteins are broken down to amino acids. Part of these amino acids are oxidatively de-aminated and the carbon skeleton enters various respiratory and carbon cycles. The

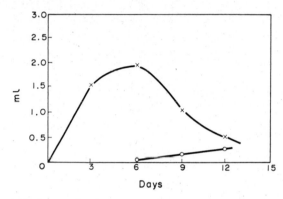

FIG. 5.13. The changes in proteinase activity of cotyledons and axis organs of soybean.

The activity of the enzyme is expressed as ml of 0·025 N KOH used in the titration. ×——× proteinase activity in the cotyledons. O——O proteinase activity in the axis organs. (From Tazakawa and Hirokawa, 1956)

ammonia formed by de-amination is detoxicated by the process of amide formation. The chief amides formed are glutamine and asparagine, depending on the plants. Not all amino acids are deaminated in this way. Part of them are utilized for the synthesis of proteins in the actually growing parts of the seedling. Soluble nitrogen already present in the seedling may also be utilized during germination. Thus Egami et al. (1957) showed that the small amounts of nitrate present in Vigna seeds disappear as germination proceeded; they also proved the existence of a nitrate reductase system in the seedlings. Yamamoto (1955) showed in the same seeds that asparagine present in the cotyledons disappears and instead appears in the hypocotyl and plumule. On the basis of these various findings the effect of feeding ammonia and sugars in protein metabolism can also be understood. In the case of feeding ammonia, protein breakdown is reduced, because external nitrogen can be used to form new protein. In the case of sugars, the sparing action is chiefly due to the provision of respiratory substrates, which would otherwise be provided by breakdown of protein and de-amination of the amino acids formed.

The mechanism of amide formation is fairly well established as far as glutamine is concerned. Glutamine is formed from glutamic acid and ammonia, in the presence of the enzyme glutamine synthetase and ATP, the reaction being energy-requiring. The precise biochemical mechanism has also been established. As far as asparagine formation is concerned the situation is less clear. However, Yamamoto showed that in *Vigna* hypocotyl, asparagine is formed if the hypocotyls are fed with aspartic acid, ammonia and ATP. The enzyme responsible for the reaction has, however, never been isolated.

In addition to enzyme systems causing the synthesis of amides, germinating seeds usually contain enzymes causing hydrolysis of the amide bond. These are glutaminase and asparaginase, whose precise function is unknown. Far more interesting are those enzymes which can transfer amino groups from the amide to some keto acid which results in amino acid formation. The enzyme responsible for glutamine formation can also cause transfer of the glutamate moiety of glutamine to some other molecule.

Another type of transfer enzyme concerned with amino acid metabolism is constituted by the transaminases. These enzymes transfer amino groups from amino acid to keto acid (see for example Webster, 1959). The presence of transaminases in a variety of seeds has been shown by Smith and Williams (1951). They found a marked increase in the activity of the transaminases transferring amino groups from alanine to α-ketoglutaric acid and from aspartic acid to α-ketoglutaric acid (Table 5.4). It is probable that the same enzymes are also

TABLE 5.4 — CHANGES IN GLUTAMIC-ASPARTIC AND GLUTAMIC-ALANINE TRANSAMINASES AND IN PROTEIN NITROGEN DURING GERMINATION OF VARIOUS PLANT SEEDS

Enzyme activity is expressed in units/embryo and protein nitrogen as mg protein/embryo. In all cases the embryo after removal of endosperm or cotyledons was used
(Smith and Williams, 1951)

Time in hours	24			48			72			96		
Seeds	Glu-Al	Glu-Asp	Protein N	Glu-Al	Glu-Asp	Protein N	Glu-Al	Glu-Asp	Protein N	Glu-Al	Glu-Asp	Protein N
Waxbean	0·03	0·10	0·11	0·11	0·16	0·16	0·15	0·28	0·26	0·17	0·48	0·32
Pea	0·03	0·06	0·09	0·17	0·13	0·12	0·28	0·18	0·15	0·37	0·23	0·16
Barley	0·08	0·09	0·03	0·14	0·26	0·04	0·24	0·56	0·08	0·42	0·88	0·11
Corn	0·09	0·10	0·08	0·17	0·17	0·11	0·20	0·28	0·16	0·44	0·53	0·32
Oats	0·05	0·14	0·02	0·10	0·18	0·02	0·17	0·44	0·04	0·20	0·67	0·07
Squash	0·06	0·04	0·08	0·07	0·06	0·08	0·08	0·13	0·08	0·13	0·43	0·09
Pumpkin	0·04	0·06	0·04	0·07	0·09	0·06	0·16	0·27	0·08	0·28	0·43	0·10

carrying out the reverse reactions. In most cases glutamic-aspartic transaminase increased more rapidly than the glutamic-alanine transaminase. However, in pea the reverse was the case, and in corn both enzymes seemed to increase at about the same rate. These authors also showed that no fixed relation existed between increase in protein in the seeds and increase in transaminase activity, and concluded that there was no evidence to show that the two processes were directly connected.

Evidence for the presence of enzymes catalysing the reverse reaction from glutamine to pyruvic or oxaloacetic acid in wheat germ was brought by Cruickshank and Isherwood (1958).

Albaum and Cohen followed the transamination reaction glutamic-oxaloacetic acid and its reverse reaction in oat embryos. The former reaction took place at three times the rate of the latter. Glutamic-oxalo acetic acid transaminase activity of the embryo expressed per unit of protein increased steadily during germination. Expressed per unit of dry weight it decreased,. probably because of the large increase in dry weight of the embryos. In contrast to the results of Smith and Williams they found good correlation between increase in protein, transaminase activity and soluble nitrogen. The changes are illustrated in Fig. 5.14.

Little information is available with regard to the importance of these reactions in amino acid and protein syntheses in the

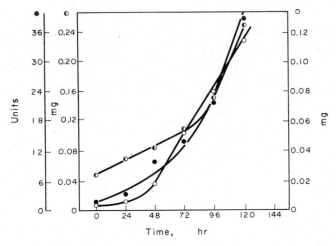

FIG. 5.14. Changes in the transaminase activity of oat embryos during germination together with changes in solubles and protein nitrogen. (After Albaum and Cohen, 1943)

◐ Protein as mg N
○ Soluble nitrogen as mg N
● Transaminase activity as units of activity

germinating seed. It must be remembered that amide formation is almost entirely absent in seedlings germinated in the light. In such seedlings, especially when the seeds are germinated in the soil or in nutrient solution, protein synthesis is from external nitrogen taken up by the seedling and from the carbon skeleton formed during photosynthesis. Undoubtedly the key processes in such seedlings are the *de novo* synthesis of amino acids and proteins and not protein breakdown and amide formation. In amino acid synthesis transamination reactions probably play an important role.

A little evidence about amino acid and protein synthesis is available. Virtanen *et al.* (1953) were able to show the synthesis of homoserine in pea seeds during the first 24 hours of germination. Homoserine was not present in the dry seed, either in the free state or in protein of the seed. Thus during germination there occurred the rapid synthesis of an entirely new amino acid. The function of this amino acid was, however, not studied. A general rise in the amino acid content of lettuce seeds during germination was shown by Klein (1955). This is shown in Table 5.5.

TABLE 5.5 — CHANGES IN THE AMINO ACID CONTENT OF
LETTUCE SEEDS DURING GERMINATION
(Klein, 1955)
The amino acid content is given as γ-amino N per gram
initial dry seeds.

Amino acid	Time of germination days			
	0	1	2	3
Alanine	5	30	80	220
Threonine	5	20	40	190
Leucine	20	20	60	280
Serine	30	30	60	250
γ-amino butyric acid	5	5	15	25
Lysine	15	5	20	40
Tryptophane	5	5	2	—
Glutathione	10	0	0	20
Aspartic acid	40	35	35	40
Glutamic acid	60	80	110	160
Asparagine	30	40	60	240
Glutamine	60	40	360	700

Changes in the free amino acid composition in the seed may be indicative of developmental changes. For example, Fine and Barton (1958) showed that both the ratio of amino acids and the absolute amounts change during after-ripening of tree peony seeds.

Protein synthesis has been shown to occur in the seed by Young *et al.* (1960). Peas were germinated in water or in solutions of compounds which specifically prevent protein synthesis, such as chloramphenicol. The formation of a number of enzymes and especially phosphatase and amylase were then followed. Those substances which specifically inhibited protein synthesis prevented the rise in phosphatase and amylase activity which normally occurs when the seeds germinate. It was concluded that the protein of the enzymes was newly formed in the seed. In further studies they were able to show that the formation of phosphatases was temperature-dependent. Moreover, both the intact cotyledons of pea seeds as well as a microsomal fraction isolated from them were able to incorporate C^{14}-labelled amino acids such as glycine into proteins, suggesting new protein formation. All this is entirely in accord with the view that protein synthesis is an important part of the process of germination.

The mechanism of amino acid and protein synthesis have hardly been studied in germinating seeds. It is therefore only possible to assume that these processes are similar to, or identical with, those known to occur in the adult plant and in other organisms. The mechanism of biosynthesis of most amino acids is today fairly well known and it may be supposed that it proceeds in the same way in the seed. It is usually assumed today that proteins are formed from amino acid without intermediary peptide formation. It appears that the amino acids are activated in the presence of ATP and are then condensed to form proteins. Protein synthesis is supposed to be under genetic control. As proteins are extremely specific in structure it has been suggested that their configuration is ultimately determined by the structure of the deoxyribonucleic acid (DNA) in the nucleus. As the bulk of protein synthesis occurs outside the nucleus in small particles, variously called ribosomes or RNA particles, it is suggested that the information present in the DNA molecule is in some way transferred to RNA molecules, which are more directly associated with protein synthesis. The exact way in which this information is transferred and the way in which RNA controls protein structure and specificity is still open to speculation.

Oota and Osawa (1954) were able to show that the microsomal particles from *Vigna* seeds and seedlings were directly concerned with protein synthesis and concluded that a relation existed between the ratio of RNA to protein in the particles and their ability to synthesize protein.

4. *Metabolism of Phosphorus-containing Compounds*

Phosphates play an extremely important role in a variety of reactions in seeds. Thus phosphate is required for the formation of the nucleic acids—which in turn are intimately connected with protein synthesis and the hereditary constitution of the plant cell. The function of phospholipids, such as lecithin, in controlling surface properties and permeability of cells is well-established today, and may be taken to be similar in the seed and seedling to that in other plant tissues. The various phosphate sugars and nucleotides are very closely linked with the energy-producing processes in the cell during germination.

Phosphorus appears in seeds primarily in the organic form and very little seems to be present as inorganic orthophosphate. Among the phosphorus-containing compounds are the nucleic acids, phospholipids, phosphate esters of sugars and nucleotides and phytin, the calcium and magnesium salt of inositol-hexaphosphoric acid. The calcium and magnesium content of the phytin molecule is variable. Thus wheat phytin contains 12 per cent calcium and 1·5 per cent magnesium, while oat phytin contains 8·3 per cent calcium and 15 per cent magnesium as well as 5·7 per cent manganese. The absolute amount of phytin is also very variable and varies not only between species but even in different varieties of the same species (Ashton and Williams, 1958). Phytin is very frequently present in many seeds and may constitute up to 80 per cent of the total phosphorus content of the seed. Some of the phosphorus-containing compounds occurring in cotton seed and the changes in them during germination are shown in Table 5.6. Because most of the phosphate is present in the bound form, orthophosphate may well be a limiting factor in many of the reactions mentioned above. For this reason the large amount of phytin present may be regarded as a store of inorganic phosphate which is liberated as germination proceeds. The liberation of this phosphate is by the enzymic hydrolysis of the phytin by a phosphatase. This enzyme is known as phytase and is probably not entirely specific for phytin and is capable of hydrolysing other phosphate ester linkages as well. As will be seen in Table 5.6,

TABLE 5.6 — CHANGES IN COMPOSITION OF THE VARIOUS
PHOSPHORUS FRACTIONS IN COTTON SEEDS, PAYMASTER
VARIETY, DURING GERMINATION
The results are given as mg phosphorus per gram dry weight
(From Ergle and Guinn, 1959)

Time of germination in days	Dry seeds	1	2	4	6
Phytin	8·61	8·49	7·15	4·00	1·97
Inorganic	0·44	0·29	1·87	4·77	7·02
Total lipid	0·71	0·81	0·87	0·50	0·85
Ester	0·32	0·41	0·40	0·58	0·42
RNA	0·12	0·11	0·15	0·25	0·39
DNA	0·11	0·11	0·12	0·21	0·44
Protein	0·11	0·10	0·16	0·28	0·26

the amount of phytin in cotton seed drops quite quickly during germination, so that after six days most of the phytin has disappeared. Concurrently, inorganic phosphate accumulates in the seeds. Similar rapid disappearance of phytin from germinating seeds has been observed in wheat, oats, peas and lettuce, as well as other seeds. In all these cases phytin is present in very considerable amounts. In cotton seeds all the phytin is present in the cotyledons (Ergle and Guinn, 1959). It is probable that in other seeds, also, a large amount of phytin is present in the storage tissues. However, Albaum and Umbreit (1943) showed that phytin is also present in the embryo of oat, disappearing rapidly during germination, but state that the endosperm contains more phosphorus than the embryo. Some of this phosphate is transported to the embryo during germination. Usually there is a good correlation between the rapidity of phytin breakdown and phytase activity

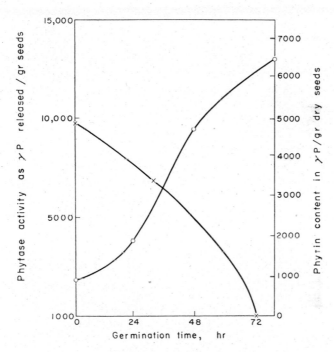

FIG. 5.15. Phytin content and phytase activity in germinating lettuce seeds (from Mayer, 1958).

Phytin content ×——× as γ phytin P per gram dry seeds
Phytase activity O——O as γ P released by gram seeds

of the seed or seedling. Such correlation is demonstrated for lettuce seeds in Fig. 5.15. In oats, phytase activity in the entire seeds seems to develop rather slowly and the total phytin content decreases to about a half in the course of a week. In wheat, on the other hand, most of the phytin has disappeared after a similar period. Peers (1953) showed that in the dry seed about 80 per cent of the phytase is present in the endosperm and only 1 per cent in the embryo. The remainder is distributed between the other parts of the seed. The enzyme prepared from wheat was characterized by stability to fairly high temperatures. The purified enzyme was less heat resistant than the crude preparation. The optimum temperature for enzyme activity was around 51°C and the pH optimum around 5·4. The data seem to be characteristic for phytase prepared from seeds.

In contrast to the large amount of data available on phytin metabolism, other aspects of phosphorus metabolism are far less clear. This is primarily due to the fact that separation of the phosphorus-containing compounds is by fairly crude fractionation procedures using precipitation as barium salts, without identification of the individual compounds present in each fraction. More refined methods using ion exchange chromatography do not yet appear to have been applied to seeds. As will be seen from Table 5.6 the changes in lipid phosphate, ester phosphate and protein phosphate in cotton seed are not very large and such changes as occur are irregular. In so far as a distinct trend is discernible, it is a slight increase in these fractions. Early work on phospholipids, and especially lecithin, in a number of seeds tend to support this, as here also an increase in phospholipids during germination was usually found. This was especially the case for seeds germinated in the light, but not all the early work is consistent on this point. In general a very large variety of sugar esters is present in seeds, including the intermediaries of glycolysis and triose phosphates, as well as nucleotides such as ATP, DPN, TPN and others such as UTP, ITP and similar and related substances. In no case has any very regular change in any of these fractions been observed. The hexose phosphate content of oat embryos rises during germination but the hexose content of the endosperm was not examined (Albaum and Umbreit 1943). The ATP content of various parts of seedlings showed an initial rise followed by a subsequent decrease. Šebesta and Šorm (1956) showed that ATP is virtually absent in the dry bean seed and is formed during imbibition. However after an initial rise in the ATP content the level decreases in all organs except the cotyledons. In contrast Albaum and Umbreit showed a steady increase in the ATP content of the embryo of oats. The changes in the total DPN and TPN content of a number of seeds was studied by Bevilacqua and Scotti (1953) and Bevilacqua (1955). They showed that the nucleotide content of the seed and seedling rises in all cases during germination. The rise in peas was much greater than in wheat or oats. In both wheat and oats the initial rise and fall occur at different times in different seeds. In wheat endosperm an initial rise up to five days, is followed by a plateau, while in peas at this time a marked drop of nucleotide content in the cotyledons is observed (Fig. 5.16). The findings with regard to ATP and DPN as well as those for sugar phosphates clearly indicate that

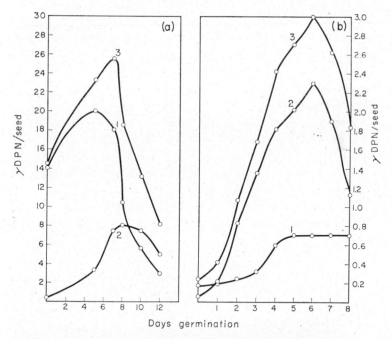

Fig. 5.16. Changes in pyridine nucleotides in seeds of *Pisum sativum* and *Triticum vulgare* during germination (Bevilacqua, 1955).

(a) – *Pisum sativum* ; (b) – *Triticum vulgare*
1 – endosperm or cotyledons; 2 – embryo; 3 – whole seed

these substances are being rapidly metabolized in the seed. This would be in accord with their function in all energy-requiring processes which occur during germination.

Of the enzymes concerned with the metabolism of nucleo-tides and sugar phosphates, only a few have been investigated to any extent in seeds. In general, most attention has been paid to general phosphatase activity on the one hand, and to the phosphokinases responsible for phosphate transfer, on the other hand. Thus in many seeds a glycerophosphatase has been shown to be present which differs from phytase both in speci-ficity, pH optimum and heat resistance. An active ATP-ase has been shown to occur in mitochondria prepared from let-tuce seeds at various stages of germination. Mitochondria from a number of seeds have also been shown to have systems carrying out oxidative phosphorylation. This will be discussed in the section on respiration. Among the phosphokinases, pyru-vic kinase appears to be of a very widespread occurrence. The same seeds usually contain a number of different phos-phatases which are able to hydrolyse phosphoenol pyruvate. The general increase in phosphatase activity has already been shown in Fig. 5.11.

Another important phosphorus-containing fraction which has been studied in some detail is that of the nucleic acids. Nucleic acids are polymers of nucleotides, which in turn are composed of purine and pyrimidine bases, ribose or desoxy-ribose and phosphate. When the nucleic acids are linked to protein they constitute nucleo-proteins. In Fig. 5.6 the chang-es in the total nucleic acid content of various parts of *Vigna* seeds are shown. These indicate a redistribution of nucleic acids between various parts of the seedling. In Table 5.6 it will be seen that in cotton seeds both RNA and DNA content rise during germination, the total increasing by a factor of three to four. This might be expected if there is an increase in cell number during germination and no storage nucleic acids are present. A good correlation between cell number and nucleic acid content was, in fact, found in beans by Maroti (1957), who followed both the increase in cell number and

TABLE 5.7 — CHANGES IN CELL NUMBER AND NUCLEIC ACID CONTENT OF
BEANS DURING GERMINATION
The numbers are per plant organ
(Compiled from Maroti, 1957)

Time	Total number of cells		γ RNA.P.		γ DNA.P.	
	Root $\times$ 10^3	Shoot $\times$ 10^3	Root	Shoot	Root	Shoot
6 h	145	66	6·6	1·9	3·0	0·1
1 day	188	80	4·2	1·1	0·9	0·9
2 days	291	104	6·5	4·4	3·3	1·1
3 days	654	193	20·5	4·3	11·0	3·1
4 days	1397	265	26·1	4·3	10·1	4·9
5 days	1316	281	29·8	7·7	30·5	4·9
6 days	1451	431	39·5	12·9	28·3	5·8
7 days	1325	808	—	15·9	—	7·0
8 days	1298	1383	58·6	18·6	13·1	11·0
9 days	1324	1595	79·1	24·0	3·2	7·5

the RNA and DNA content of roots and shoots (Table 5.7).
In the root both cell number and nucleic acid content rose
more quickly than in the shoot and it was concluded that the
root has greater synthetic activity than the shoot. However,
when the cotyledons had been depleted of reserve materials
and the seedlings kept in the dark, the shoot became increas-
ingly active and apparently materials were transported to it.
The complexity of nucleic acid metabolism in the seeds is
emphasized by the continuation of the work on *Vigna* by
Oota and Takata (1959). These authors separated the RNA
from the beans electrophoretically and were able to show at
least four different components. They conclude that two of
these can be identified respectively as functional and transportable RNA. The functional RNA is linked to protein and
is involved in protein synthesis. The transportable RNA seems
to be separate from the proteins and constitutes those RNA
molecules which are transported through the seedling. The
other components observed by Oota and Takata are regarded
by them as intermediaries in the conversion of functional to

transportable RNA or vice versa. Thus any observations on the nucleic acid metabolism of the seed and seedling must take into account this special feature of different nucleic acid components and their conversion, as well as the actual movement of the nucleic acid through the seedling. The significance of nucleic acid in protein synthesis has already been mentioned. The site of RNA synthesis is presumed to be primarily in the ribonucleic acid particles. Evidence of this in pea seedlings was brought by Webster (1957) who showed incorporation of carbon from various precursors, including amino acids and purines into the nucleic acid of these particles.

III. Respiration

1. *Gaseous Exchange*

Germination is an energy-requiring process and is therefore dependent on the respiration of the seed. In the following account we will discuss respiration both from the point of view of overall gas exchange, and from the point of view of biochemical mechanism.

In dry seeds it is almost impossible to measure either oxygen uptake or carbon dioxide output. There can be no doubt whatsoever that such gas exchange as exists in the dry seeds is at an extremely low level. The problem of measuring the gas exchange of dry seeds is further complicated by the fact that most seeds are to some extent contaminated, both on the seed coat and frequently also between the seed coat and the seed, with bacteria and fungi. These micro-organisms also have some kind of respiration and it is more than likely that some, if not all, the gas exchange measured in dry seeds is in fact due to the contaminating micro-organism. For example, Rose (1915) found that out of a hundred species examined, more than half were infected by fungi. When large bulks of seed are kept, as in grain silos, there is an appreciable heat production, resulting in a rise in temperature. The heat produced is presumably also due, at any rate in part, to the respiration of micro-organisms. It is worth pointing out that it is very difficult effectively to sterilize seeds and surface sterilization may, in fact, not be sufficient. In the few established cases of very long-lived seeds, i.e. *Nelumbo*, it is difficult to understand how any form of respiration could have been maintained in the seeds for such long periods of time without entirely depleting the storage materials of the seeds, thus impairing their viability. Despite these reservations there is a certain amount of information which shows the existence of gas exchange in dry seeds. The level of gas exchange of dry seeds is very definitely dependent on the moisture content of the seeds and rises as the latter rises. Thus Bailey (1921) showed that in *Zea mays* seeds the output of carbon dioxide rose from 0·7 mg per hundred gram dry weight during twenty-four hours, when the seeds had a moisture content of 11 per cent, to about 60 mg when the moisture content was 18 per cent. Similar increases in carbon dioxide output with increasing moisture content have been shown for sorghum, wheat and rice. The steepness of the rise in carbon dioxide output with increasing moisture content differs in different seeds.

As seeds take up water there is generally speaking a marked increase in their gas exchange. However, it has been shown that on moistening seeds with water there is an immediate gas release. This gas release seems to be a purely physical process not peculiar to seeds and involving the liberation of gas which is supposed to be colloidally absorbed within the seeds (Haber and Brassington, 1959). This observation obviously complicates any interpretation of the gas exchange of seeds immediately after they are placed in water.

Another complicating factor in respiration measurements of seeds is the presence of the seed coat. The problem of the permeability of the seed coat has already been discussed (see Chapters 3 and 4). This factor also affects the course of respiration directly. Thus in *Lathyrus* seeds the respiration consists of several phases, one of which is a plateau when the rate of respiration remains constant. If the testa is removed these

phases are virtually absent and the rate of respiration rises
steadily with time (Stiles, 1935).

The changes in the oxygen uptake (Q_{O_2}), carbon dioxide
output (Q_{CO_2}) and the respiratory quotient $RQ = Q_{CO_2}/Q_O$

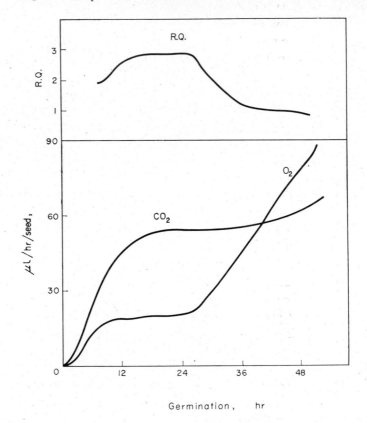

FIG. 5.17. The rate of respiration and the respiratory quotient (RQ)
of pea seeds during the first 50 hours of germination. Temp. 25°C.
(Spragg and Yemm, 1959)

of different seeds are illustrated in Figs. 5.17, 5.18 and
5.19. From these figures the general increase in both Q_{O_2} and
Q_{CO_2} with time are evident. It will also be noted that in
wheat and flax the rise of both Q_{O_2} and Q_{CO_2} is more or less
uniform during the early stages of respiration and only falls,

in the case of flax, when the seedlings become older (Fig. 5.18). In peas, on the other hand, the plateau mentioned above can be clearly seen (Fig. 5.17). This plateau appears to end at about the time when the seed envelope is broken and free gas exchange, without limitation by membranes, becomes possible (Spragg and Yemm, 1959). However, it is worth noting

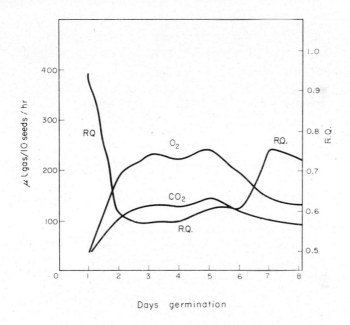

FIG. 5.18. The rate of respiration and the RQ of germinating flax seeds. Temp. 20°C. (Halvorson, 1956)

that the plateau for Q_{O_2} and Q_{CO_2} ends at different times. Moreover, the removal of the testa, although raising the total oxygen uptake did not result in an abolishing of the plateau. In all three cases it is important to note that although both oxygen uptake and carbon dioxide output rise with time, they rise at quite different rates. As a result the RQ during the early stages of germination shows very large variations. These point to very profound changes in the substrates used for respiration, a point which will be returned to later.

The RQ is dependent on the state of oxidation of the substrate oxidized. Highly oxidized substrates such as organic acids result in a RQ of between 1·0 and 1·5 while fats give RQs of the order of 0·7–0·8. An RQ of 1·0 is characteristically obtained if the substrate respired is a carbohydrate. Further, the RQ obtained depends on the extent to which there is genuine respiration and to what extent fermentative processes

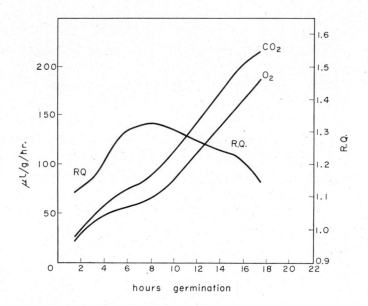

Fig. 5.19. The rate of respiration and the RQ of germinating wheat seeds. Temp. 20°C. (Levari, 1960).

occur. In seeds with very compact tissues fermentation usually occurs initially, and only when oxygen penetrates into the tissues does respiration proper begin. In these cases there will initially be a marked carbon dioxide output due to glycolysis and only a slight oxygen uptake resulting in a very high RQ although the substrate broken down may be a carbohydrate.

An indication of the difference in Q_{O_2} and RQ in different organs of the same seed is shown in Fig. 5.20. This shows that while embryo, cotyledons and endosperm all show an initial rise in Q_{O_2} the rate of oxygen uptake of the endosperm continues to rise till about six days and then falls again as the endosperm finally disintegrates. In contrast cotyledons and

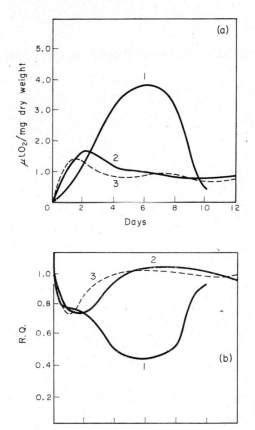

Fig. 5.20. Oxygen uptake and RQ values of various parts of the castor bean seed during germination. (Yamada, 1955)
1 – endosperm; 2 – cotyledons; 3 – embryo axis (or hypocotyl).
(a) – oxygen uptake; (b) – RQ values

embryo rapidly attain a steady state, if respiration is expressed per gram dry weight. These experiments were made using tissue slices of the endosperm and the hypocotyl, a method which will presumably alter the oxygen uptake of the organs as compared to their normal behaviour, both qualitatively and possibly also quantitatively. It is interesting to note that the RQ of the endosperm also shows very marked changes while in the embryo relatively small changes occur.

The rate of respiration of seeds is influenced, as is respiration in general, by a number of external factors. Among those which must be recalled are the moisture content of the seeds, which has already been discussed, temperature, oxygen and carbon dioxide content of the external atmosphere and light. Each of these factors is liable to affect respiration somewhat differently depending on the precise stage of germination during which its influence is examined.

Generally, a rise in temperature causes an increase in the rate of respiration in the seeds. However, the early experiments of Fernandes (1923) showed that the oxygen uptake at different temperatures depends not only on the actual temperature, but also on the length of time the seeds are exposed to this temperature. In other words, in studiyng the effect of temperature, the time factor must also be taken into account. The respiration of flax seeds germinated at different temperatures is shown in Fig. 5.21 (a). It has also been shown that the effect of temperature depends on the presence or absence of the testa. For example in peas whose testa had been removed, an increase in temperature raised the oxygen uptake much more than in intact pea seeds [Fig. 5.21 (b)] (Spragg and Yemm 1959). Thus temperature can only effect respiration provided oxygen can freely diffuse to the respiring tissue. If oxygen diffusion is limited, an increase in temperature will have relatively little effect.

An increase in the oxygen tension can also increase the rate of respiration of seeds. Many examples of this are available. In many cases, however, this only applies to oxygen concentrations below 20 per cent, maximum rates being reached at this value. This was shown to be the case for *Triticum spelta* and *Brassica rapa* (Reuhl, 1936). However in *Linum usitatissimum*, in the early stages of germination when the root had just emerged, oxygen uptake had not yet reached a steady state at 20 per cent oxygen. In most cases, in the later stages of germination, when the roots were 1–3 cm long, steady states of oxygen uptake were obtained at 20 per cent oxygen. For peas it was recently shown that the effect of raised oxygen concentration depended on the stage of germination chosen for study. Respiration in pure oxygen was appreciably higher than in air, when the seeds had been germinated up to 36 hours. Up to this time there was a steady rise in the percentage increase of

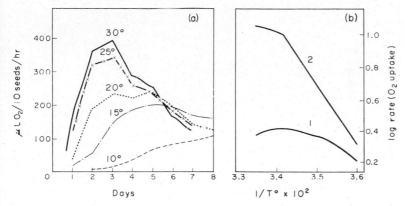

FIG. 5.21. Effect of temperature on the respiration of germinating seeds.
(a) Oxygen absorption of flax seeds during germination at different temperatures (Halvorson, 1956).
(b) The respiration of pea seeds with or without testas. Seeds were germinated for 24 hours.
1 – whole seeds; 2 – seeds with their testa removed (Spragg and Yemm, 1959).

respiration in oxygen as compared to air. After 36 hours the percentage increase dropped again and at 48 hours the rate of respiration in air and in oxygen were about equal.

Light has often been claimed to affect respiration of various tissues. Mostly the observed results have been obtained using white light. In view of the profound effect of light on germination previously discussed, its effect on the respiration of seeds is of some interest. It may be expected that, in so far as the total germination is increased by light, the respiration also would be increased. However, it is possible that there are other effects. The influence of red and far-red light on respiration has been studied both by Leopold and Guernsey (1954) and by Evenari, Neumann and Klein (1955). Both groups used light-sensitive lettuce seeds but their experimental techniques differed. The essential factor common to the results of both groups is that red light resulted in a raised respiration of the seeds, shortly after illumination had been applied and before any visible germination could be observed. Far-red light reversed this increase in oxygen uptake. Far-red light alone reduced the oxygen uptake. Evenari *et al.* found differences in the exact behaviour of the seeds in response to red and far-red light depending on the storage period of the seeds. Changes in the RQ, in response to illumination, were also observed, apparently because red light raised the Q_{CO_2} but did not affect the Q_{O_2} while infra-red light depressed the Q_{O_2} but did not affect the Q_{CO_2}. Leopold and Guernsey studied the effect of red light and far-red light on the respiration of other tissues known to be affected by the red far-red mechanism. They found that there was a good correspondence between the effect of light and the response of respiration. When red light stimulated a process, stimulation of respiration resulted and a depression in respiration was observed when red light had an inhibitory effect. In every case far-red light reversed the effect of red light. More recently a much more direct effect of red and far-red light has been demonstrated by Gordon and Surrey (1960). They were able to show that oxidative phosphorylation by mitochondria, isolated from oat coleoptiles, depended on the treatment of the intact coleoptiles with red or far-red light. However, the results were not consistent, red light sometimes depressing and sometimes stimulating oxidative phosphorylation, depending on the precise period at which it was given and the age of the coleoptiles.

The respiration of entire seeds is sensitive to externally applied respiratory inhibitors. Such inhibitors, provided they penetrate into the seeds and are not metabolized in it, act in much the same way as they act in other tissues. The relation between inhibitory action and the process of germination as a whole will be discussed later, as will the effect of other germination inhibitors as well as stimulators.

2. *Biochemical Aspects of Respiration*

Respiration is that process in which the substrate is oxidized through a series of steps, with the participation of oxygen as the final electron acceptor. All other processes in which oxygen does not participate are not strictly respiration. They are frequently termed anaerobic respiration, but the term fermentation is a more correct one to use.

The mechanism of respiration and fermentation can be extremely varied. The best-established mechanism of respiration is that of glycolysis in which the substrate is broken down to the level of pyruvate, followed by the oxidation of the pyruvate in the tricarboxylic acid or Krebs cycle. An alternative mechanism of oxidation is the direct oxidation of glucose phosphate leading to metabolism of the pentose cycle. An additional by-pass of the oxidative process is the glyoxylic acid cycle. Fermentative processes using carbohydrate as the substrate go via the glycolytic pathway. The pyruvate formed is either decarboxylated, leading to carbon dioxide formation and the resultant acetyl derivative is reduced to alcohol, or the pyruvate is directly reduced, leading to the formation of lactic acid. Although many other fermentation processes are known in micro-organisms, no information is available about their existence in plant tissues.

The main processes known with certainty to yield energy available to the organism are oxidative phosphorylation linked to the electron transport occurring in the Krebs cycle, and phosphorylation during glycolysis. The electron transport associated with the Krebs cycle is via a chain of enzymes consisting usually of dehydrogenases, flavoproteins and cytochromes, ending in cytochrome oxidase which transfers electrons directly to oxygen. Various parts of this electron-transport system are coupled to phosphorylation and the average ratio of phosphate esterified to oxygen taken up is $P/O = 3$.

Alternative electron-transport systems having different intermediaries and a different terminal oxidase are known.

Among those that may function in plant tissues are the gluta-thione-ascorbic acid-ascorbic acid oxidase system, the phenol-phenolase system and the glycolic acid-glycolic acid oxidase systems. In none of the alternative electron transport systems mentioned is there any evidence to show coupling to phos-phorylation and their precise function is unclear.

We will attempt in the following to bring such evidence as is available to show the existence of these various pathways in germinating seeds and try and evaluate the relative impor-tance of them during germination.

The existence of glycolysis may be reasonably assumed in seeds, as its presence in many plant tissues has been shown (see review by Stumpf, 1952). But conclusive evidence for its occurrence has only been brought in a few cases. Probably the clearest proof for the existence of glycolysis is for pea seeds, in which all the necessary enzymes seem to occur (Hatch and Turner, 1958). These workers prepared extracts from peas which were capable of catalysing the glycolytic reaction. The ability to carry out glycolysis was not followed through dif-ferent stages of germination. However, the fact that peas can, during germination, accumulate alcohol or lactic acid or both is well established. It may therefore be supposed that during the early stages of germination, as well as in the imbibed seeds, glycolysis occurs. Hatch and Turner were unable to find any evidence for any other mechanism by which substances such as glucose or glucose phosphate were broken down and, on the contrary, in their extract there was good correspondence be-tween the expected formation of alcohol and carbon dioxide if only glycolysis occurred, and the actual observed values (Table 5.8). Further evidence for the existence of glycolysis has been brought by showing that extracts of both peas and pea seedlings take up inorganic phosphate from the medium by glycolytic phosphorylation (Mayer, 1959; Mayer and Mapson, 1962). This showed that the essential phosphate trans-ferring system, linked to glycolytic reactions, functions in pea extracts. The accumulation of alcohol and lactic acid as end products does not usually occur in seeds which are germinated

TABLE 5.8 — GLYCOLYTIC CONVERSION OF VARIOUS SUBSTRATES BY PEA EXTRACTS
(After Hatch and Turner, 1958)

Substrates	CO_2 production μ moles		Alcohol production μ moles	
	Calculated	Found	Calculated	Found
Fructose 21 μmoles + Fructose diphosphate 2·5 μmoles	47	40	47	41
Glucose phosphate 25·2 μmoles	50·4	46	50·4	49
Fructose diphosphate 25 μmoles	50	41	50	44

under conditions of good aeration. In a number of seeds alcohol will accumulate if they are germinated under certain conditions, such as poor aeration for lettuce (Leggat, 1948), or high temperatures in the case of bean cotyledons (Oota et al., 1956). In other seeds, for example wheat, certain of the enzymes participating in glycolysis have been shown to be present. In Zea mays seedlings the formation of alcohol dehydrogenase was found to be increased by anaerobic conditions, both in the scutellum and the embryonic axis. Apparently the substance causing indirect alcohol dehydrogenase formation was acetaldehyde (Hageman and Flesher, 1960). The same conditions which caused an increase in dehydrogenase were accompanied by a drop in cytochrome oxidase activity in the seedlings. In none of these cases has the complete glycolytic system been demonstrated, but alcohol formation and the presence of the necessary enzymes is indicative of its presence.

Evidence for the existence of the pentose phosphate cycle in seeds is extremely scant. Glucose-6-phosphate and phosphogluconate dehydrogenases have been shown in a number of seeds, including wheat and lettuce. Other enzymes of the pentose phosphate cycle are present in various plant tissues. Convincing evidence for the existence of the pentose phosphate cycle have been brought for mung beans (Chakravorty and

TABLE 5.9 — CHANGES IN THE ACTIVITY OF GLUCOSE-6-PHOSPHATE AND
PHOSPHOGLUCONATE DEHYDROGENASE DURING GERMINATION OF MUNG
BEANS (*Phaseolus radiatus*).
Results are given as units/ml crude extract for total activity and
units/mg protein for specific activity
(After Chakravorty and Burma, 1959)

Age of seedling in hours	Activity of glucose-6-phosphate dehydrogenase		Activity of phosphogluconate dehydrogenase	
	Total	Specific	Total	Specific
24	0·50	0·05	0·50	0·05
48	0·50	0·06	0·40	0·05
72	0·27	0·05	0·25	0·05
96	0·08	0·02	0·11	0·03

Burma, 1959). The presence of all the enzymes necessary for
the oxidation of glucose-6-phosphate to ribulose phosphate
and further conversion of the latter was proved. In addition,
evidence was brought that this system can only be operative
in seedlings up to three days, as the vital glucose-6-phosphate
dehydrogenase almost disappears after the third day (Table
5.9). Of course the existence of enzymes carrying out a process or a reaction does not yet prove the presence of this reaction *in vivo*, but it seems very likely from other information on
the metabolism of seeds that such a pentose phosphate cycle
may be functioning. Nothing is known of the possible energy
production, if any, associated with this pathway.

The tricarboxylic acid cycle is, generally speaking, very
widespread in its occurrence. Three criteria have been proposed
as evidence for its functioning. These criteria demand (a) proof
of the existence of the complete cycle, (b) that the cycle can be
entered at any point, and (c) that any substance which is an intermediary of the cycle can be used as a substrate. According
to these criteria there is little or no evidence to show that the
tricarboxylic acid cycle functions in seeds. However, there are
many other approaches to studying this acid cycle. Such
methods include showing of the existence of various enzymes
of the cycle, its linkage to oxidative phosphorylation, oxygen

uptake associated with specific substrates and so on. Many of such approaches have been used for seeds.

The most usual way of adducing evidence for the existence of tricarboxylic acid cycle is by the use of isolated mitochondria. Such mitochondria are then tested for their ability to oxidize the various intermediaries of the cycle and to show that oxidation of the intermediaries is coupled to oxidative phosphorylation. The ability of mitochondria prepared from seeds to oxidize tricarboxylic acid cycle intermediaries has been shown for peas, peanuts, mung beans, lettuce, castor beans and others. The linkage of oxidation to phosphorylation is experimentally more difficult to show. The P/O ratios obtained are usually lower than those obtained for animal material. In Table 5.10 are shown the changes in the oxidative ability of mitochondria isolated from lupin seeds and seedlings at different ages. It will be seen that the oxidative ability of the mitochondria generally rises as the age of the seedling increases but not at equal rates for different substrates. Clear proof for a cyclic process has been brought for castor beans (Neal and Beevers, 1960). If slices of castor bean endosperm were fed with labelled pyruvate, it was possible to follow the fate of the various carbon atoms of the pyruvate. Pyruvate labelled with C^{14} in the one position gave rise to labelled carbon dioxide rapidly and quantitatively. When, however, the label was in the

TABLE 5.10 — ABILITY OF MITOCHONDRIA ISOLATED FROM LUPIN SEEDS TO
OXIDIZE VARIOUS SUBSTRATES
Results as μl O_2/hr/mg N, (Conn and Young, 1957)

Age of seedling	Oxygen uptake			
Substrate	12 hours	2 days	4 days	10 days
Succinate	19·0	70·0	167·0	208·0
α-Ketoglutarate	49·0	100·0	122·0	118·0
Malate	42·0	46·0	72·0	72·0
Citrate	30·0	25·0	74·0	90·0
Glutamate	3·0	65·0	68·0	91·0
Endogenous	4·0	4·0	9·0	5·0

two or three position the C^{14} label spread slowly through various cell constituents and the labelled carbon was liberated only very slowly as carbon dioxide. This indicated that the C-1 carbon is removed by immediate decarboxylation while the C-2 and C-3 carbons are being cycled and only slowly take part in decarboxylation reactions. In corn mesocotyls the C-2 and C-3 carbons were not released at all as carbon dioxide. Instead they appeared in various amino acids and in protein, presumably by being diverted at various stages of the tricarboxylic acid cycle.

The behaviour of mitochondria does not necessarily reflect the behaviour of the entire cell. It is well known today that the ability of isolated mitochondria to oxidize a certain substrate depends very largely on the method of isolation. The tonicity of the medium, the presence of certain ions in the isolation medium and other preparative details all play a part in determining the activity of mitochondria, as does the exact composition of the reaction mixture in which their oxidative capacity is being tested. Further, it must be remembered that the ability of mitochondria to oxidize some of the substrates of the tricarboxylic acid cycle does not prove the existence of the cycle, because of possible by-passes of it. Additional evidence is normally necessary. Such evidence may come from tracer studies, the use of inhibitors and the chromatographic identification of the intermediaries which occur during the cyclic process. Even if the existence of a complete cycle is established, its overall contribution to metabolism of the tissue will be determined by the rate of oxidation of that substrate which is least readily utilized and is rate-limiting.

An important pathway by which pyruvate may be metabolized is via the glyoxylate cycle. The relation of this metabolic pathway to the tricarboxylic acid cycle on the one hand, and to the conversion of fats to carbohydrates on the other hand, is shown in Fig. 5.22. The important feature of this cycle is that acetyl CoA which is formed from pyruvate condenses with glyoxylic acid to form malate, in the presence of the enzyme malate synthetase and ATP. The glyoxylate arises from isocitric acid, with the concomitant formation of succinic acid by the action of isocitritase. The succinic acid can enter the tricarboxylic acid cycle, as does the malate which is formed from the glyoxylate. Evidence for the operation of this pathway has been brought for both *Arachis* and *Ricinus communis* seeds (Kornberg and Beevers, 1957; Marcus and Velasco,

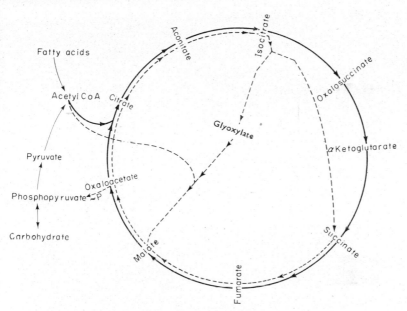

FIG. 5.22. The relation between the tricarboxylic acid cycle (solid line) and the glyoxylate cycle (broken line) and the possible way of conversion of fat to carbohydrate.

1960; Yamamoto and Beevers, 1960). It appears that in both castor beans and peanuts the chief function of the glyoxylic acid cycle is the conversion of fats to carbohydrates. This conversion is effected via the oxidation of the fatty acids by β-oxidation (as previously mentioned) and formation of acetyl CoA, which is in turn converted to malate. The malate is converted to carbohydrates via oxaloacetate and phosphopyruvate, a process which requires ATP. At the same time it seems that α-ketoglutarate in the tricarboxylic acid cycle can be completely by-passed due to the operation of the glyoxylic acid cycle (Marcus and Velasco, 1960). An alternative point of entry of two carbon units into the respiratory mechanism is provided by this means. The activity of the enzymes involved in this cycle has been shown to increase during germination of *Arachis* seeds, as shown in Table 5.11. It will be seen

TABLE 5.11 — CHANGES IN ACTIVITY OF THE GLYOXYLATE
CYCLE DURING GERMINATION OF PEANUTS
Enzyme activity given as μmoles glyoxylate formed
(isocitritase) or removed (malate synthetase)
(From Marcus and Velasco, 1960)

Time in days	Length of radicle (mm)	Enzyme activity	
		Isocitritase	Malate synthetase
0		—	
1	0	0	0
2	2	7·9	10·2
3	16	20·4	24·6
4	35	30·0	50·2
5	55	42·3	69·4

that both malate synthetase and isocitritase increase during germination and are completely absent in the dry seed. All the enzymes of the glyoxylic acid cycle are located in the mitochondria.

In all those cases where a substrate is oxidized during respiration, the primary step is the removal of hydrogen from the substrate molecule by the action of dehydrogenases. These dehydrogenases can transfer hydrogen or electrons further by a number of alternative mechanisms. The immediate hydrogen acceptor may be one of the co-enzymes DPN or TPN or, as in the case of the oxidation of succinic acid, it may be a flavoprotein. TPN is the chief hydrogen acceptor in the case of glucose phosphate dehydrogenase, while most of the dehydrogenases of the tricarboxylic acid cycle are linked to DPN. In the usual electron transport chain which is associated with this cycle, the final electron acceptor is oxygen and, as already mentioned, the intermediaries in electron transport are the cytochromes and possibly other co-factors. Although theoretically more than three steps in the electron transport chain could lead to ATP formation by phosphorylation, in fact it apparently occurs only in three of them.

The general activity of dehydrogenases has been surveyed in a number of seeds including cereals, lupin, lettuce, beans, peas as well as others. In early experiments the activity of dehydrogenases was assayed by following the decoloration of some suitable dye, such as methylene blue, under anaerobic conditions. These experiments showed, in general, the presence of a variety of dehydrogenases in seeds and seedlings of many plants. No very clear conclusion emerges from these experiments. Certainly it does not seem to be a general conclusion that dehydrogenase activity increases uniformly as germination proceeds. Rather it seems that some dehydrogenases increase in activity, others remain at a steady level while yet others actually seem to decrease in activity. The complexity of the situation is illustrated in Table 5.12. It will be seen that not

TABLE 5.12 — CHANGES IN THE ACTIVITY OF VARIOUS DEHYDROGENASES IN DIFFERENT PARTS OF THE SEEDLINGS OF *Vigna sesquipedalis*
Enzyme activity is given in relative units calculated on a dry weight basis for each organ
(Compiled from Oota, Yamamoto and Fujii, 1954)

Time of germination	Citric Dehydrogenase				Malic Dehydrogenase				α-Ketoglutaric dehydrogenase			
	a	b	c	d	a	b	c	d	a	b	c	d
0	—	—	—	0	—	—	—	15	—	—	—	8
1	53	25	15	0	16	53	79	85	126	26	146	45
2	103	57	25	0	32	60	35	53	82	15	83	25
3	134	72	16	0	46	43	20	22	29	8	44	7
4	155	60	24	0	33	36	23	8	9	10	26	5
5	160	57	21	0	35	43	28	3	7	13	6	0
6	180	50	21	0	35	45	21	0	7	12	0	0

a – plumule   b – hypocotyls   c – radicle   d – cotyledons only do different dehydrogenases change in activity at different rates but also that the same dehydrogenase changes its eactivity quite differently in different organs of the same seedling. Oota *et al.* (1956) concluded that the oxidative mechanisms in the cotyledons differ from those in other parts of the seedling which are growing actively. They also assume that the tricarboxylic acid cycle is absent from the cotyledons, as citric dehydrogenase is not functioning in them. Although the dehydrogenases differ widely, nevertheless the presence of dehydrogenase is frequently taken as a proof for the viability of a certain seed. Viability tests on dry seeds are usually based on colour reactions due to the action of dehydrogenases.

Practically nothing is known about the oxidation of the hydrogen acceptors of the dehydrogenases. There is evidence for the existence of cytochrome-*c* oxidases in peas, *Zea mays*, barley, lettuce and *Ricinus* seeds. In the latter seeds cytochrome reductase was also shown to be present. Discrepancies seem to exist between observations made with crude homogenates as opposed to isolated mitochondria. Mitochondria isolated from lupin seedlings were able to carry out oxidativephosphorylation, a P/O of over 3 being obtained when α-ketoglutarate was oxidized (Conn and Young, 1957). These values have until now only been obtained in tissues containing cytochrome-*c* oxidase as the final electron transferring enzyme, and in which electron transport is via the normal cytochrome chain. There is therefore a good deal of indication that the cytochrome system may function in seeds but rigorous proof seems to be lacking.

So far, attempts to show oxidative phosphorylation by mitochondria prepared from dry seeds have been unsuccessful. This is in agreement with the observation that the tricarboxylic acid cycle seems to be operating at best sluggishly in dry seeds or mitochondria isolated from them. When the seedlings are a little older, both the tricarboxylic acid cycle operates more vigorously and it becomes possible to show oxidative phosphorylation by the mitochondria, e.g. in lettuce, lupin, peas and soybeans (Poljakoff-Mayber, unpublished;

Conn and Young, 1957; Switzer and Smith, 1957; Laties, 1957). The first to succeed in showing oxidative phosphorylation in plant tissues was Millerd (1951) who used mung bean seedlings. The P/O ratios obtained with most of the plant material was of the order of 1–2 for succinate, and variable between 1 and 3 for α-ketoglutarate, but values of near to 4 were obtained when the oxidation was confined to one step. Probably part of the difficulty in showing oxidative phosphorylation in mitochondria from seeds and seedlings is due to the presence of highly active ATP-ases in the mitochondria, for example in lettuce, peas and other seeds.

In addition to the conventional electron-transport system associated with the mitochondria there is scattered evidence for alternative electron-transport systems. In wheat germ it has been shown that reduced TPN can be oxidized by molecular oxygen in the presence of two enzymes, one of which is peroxidase (Conn et al., 1952). This is of interest in view of the evidence previously recalled for the formation of $TPNH_2$ during the pentose phosphate shunt. The mechanism of oxidation of $TPNH_2$ and the possible energy release during its oxidation are at present in dispute. A soluble DPNH-oxidase has been shown to be present in dry lettuce seeds as well as in lettuce seedlings. This latter enzyme may be a phenolase (Mayer, 1959). A complete alternative electron transport system has been shown to be present in pea seedlings. This system consists of a dehydrogenase, TPN, glutathione, and ascorbic acid in the presence of ascorbic acid oxidase. In dry seeds the system did not seem to function because ascorbic acid oxidase was absent (Mapson and Moustafa, 1957). This system was able to mediate 20–25 per cent of the total respiration of the young seedling and apparently was functioning in this way in the intact seedling after about three days of germination.

Many other oxidative enzymes are known to occur in plant tissues. Among those which have been shown to be present in seeds are catalase, peroxidase, lipoxidase as well as phenolase. The changes in these enzyme systems with germination has been followed in many cases. Unfortunately nothing is known as to how these enzymes are integrated into multi-enzyme systems and how, if at all, they are related to respiration. Phenolase has been frequently supposed to function as a terminal oxidase which can oxidize reduced co-enzymes in the presence of a phenol. There is some indication of it so functioning, for example in lettuce seeds, but this is by no means certain. Peroxidases and possibly also catalases may be connected in some way to the direct oxidation of flavoproteins by molecular oxygen, but again convincing evidence is lacking.

In conclusion it may be stated that our knowledge of the respiratory systems in seeds is still extremely incomplete. Although some parts of some of the respiratory pathways are known, there is good reason to believe that these pathways are in no way fixed and immutable, and they may change both during germination and as a result of changing environmental factors.

BIBLIOGRAPHY

ALBAUM, H. G. and COHEN, P. P. (1943) *J. Biol. Chem.* **149,** 19.

ALBAUM, H. G. and UMBREIT, W. W. (1943) *Amer. J. Bot.* **30,** 553.

ASHTON, W. M. and WILLIAMS, P. C. (1958) *J. Sci. Food. and Agr.* **9,** 505.

BAILEY, C. H. (1921) Univ. of Minnesota. Agr. Expt. Stat. Tech. Bull, 3.

BEVILACQUA, L. R. and SCOTTI, R. (1953) *Acad. Ligure di Sci. e Let.* **X,** 1.

BEVILACQUA, L. R. (1955) *Rend. Acad. Naz. Lincei ser.* VII, **18,** 214.

BOATMAN, S. G. and CROMBIE, W. M. (1958) *J. Exp. Bot.* **9,** 52.

CHAKRAVORTY, M. and BURMA, D. P. (1959) *Biochem. J.* **73,** 48.

CHIBNALL, A. C. (1939) *Protein Metabolism in the Plant.* Yale Univ. Press.

CONN, E. E., KRAEMER, L. M., PEI-NAN LIN and VENNESLAND, B. (1952) *J. Biol. Chem.* **194,** 143.

CONN, E. E. and YOUNG, L. C. T. (1957) *J. Biol. Chem.* **226,** 23.

CRUICKSHANK, D. H. and ISHERWOOD, F. A. (1958) *Biochem. J.* **69,** 189.

DAMODARAN, M., RAMASHWAMY, R., VENKATESAN, T. R., MAHADEVAN, S. and RANDAS, K. (1946) *Proc. Ind. Acad. Sci.* B **23,** 86.

DUPERON, R. (1958) *C. R. Acad., Sci. Paris* **246,** 298.

DUPERON, R. (1960) *C. R. Acad., Sci. Paris* **251,** 260.

DURE, L. S. (1960) *Plant Phys.* **35,** 925.

EDELMAN, J., SHIBKO, S. I. and KEYS, A. J. (1959) *J. Exp. Bot.* **10,** 178.

EGAMI, F., OHMACHI, K., IIDA, K. and TANIGUCHI, S. (1957) *Biochimia,* **22,** 122.

ERGLE, D. R. and GUINN, G. (1959) *Plant Phys.* **34,** 476.

EVENARI, M., NEUMANN, G. and KLEIN, S. (1955) *Physiol. Plant* **8**, 33.
FERNANDES, D. S. (1923) *Rec. Trav. Bot. Neerl.* **20**, 107.
FINE, J. M. and BARTON, L. V. (1958) *Cont. Boyce Thompson Inst.* **19**, 483.
FUKUI, T. and NIKUNI, Z. (1956) *J. Biochem., Japan* **43**, 33.
GORDON, S. A. and SURREY, K. (1960) *Radiation Research* **12**, 325.
HABER, A. H. and BRASSINGTON, N. (1959) *Nature, Lond.* **183**, 619.
HAGEMAN, R. H. and FLESHER, D. (1960) *Arch. Biochem. Biophys.* **87**, 203.
HALVORSON, H. (1956) *Physiol. Plant.* **9**, 412.
HARDMAN, E. E. and CROMBIE, W. M. (1958) *J. Exp. Bot.* **9**, 239.
HATCH, M. D. and TURNER, J. F. (1958) *Biochem. J.* **69**, 495.
KIRSOP, B. H. and POLLOCK, J. R. A. (1957) *European Brewery Convention*, p. 84.
KLEIN, S. (1955) Ph. D. Thesis, Jerusalem. (In Hebrew)
KORNBERG, H. L. and BEEVERS, H. (1957) *Biochim. Biophys. Acta* **26**, 531.
LATIES, G. C. (1957) *Survey of Biological Progress* III, 213.
LECHEVALLIER, D. (1960) *C. R. Acad. Sci., Paris* **250**, 2825.
LEGGATT, C. W. (1948) *Canad. J. Res.* C **26**, 194.
LEOPOLD, A. C. and GUERNSEY, F. (1954) *Physiol. Plant* **7**, 30.
LEVARI, R. (1960) Ph. D. Thesis. Jerusalem. (In Hebrew)
MCCONNELL, W. B. (1957) *Canad. J. Biochem. Physiol.* **35**, 1259.
MCLEOD, A. M. (1957) *New Phyt.* **56**, 210.
MCLEOD, A. M., TRAVIS, D. C. and WREAY, D. G. (1953) *J. Inst. Brew.* **59**, 154.
MAPSON, L. W. and MOUSTAFA, E. M. (1957) *Biochem. J.* **62**, 248.
MARCUS, A. and VELASCO, J. (1960) *J. Biol. Chem.* **235**, 563.
MAROTI, M. (1957) *Acta Biologica Academiae, Scientiarum Hungaricae*, VII, 277.
MAYER, A. M. (1958) *Enzymologia* **19**, 1.
MAYER, A. M. (1959) *Enzymologia* **20**, 13.
MAYER, A. M. (1959) *Proc. Int. Bot. Cong.* Vol. 2, 256.
MAYER, A. M. and Mapson, L. W. (1962) *J, Exp. Bot.* **12** (in press).
MILLERD, A. (1951) Ph. D. Thesis, University of Sydney, Australia.
NEAL, G. E. and BEEVERS, H. (1960) *Biochem. J.* **74**, 409.
OOTA, Y., FUJII, R. and OSAWA, S. (1953) *J. Biochem., Tokyo* **40**, 649.
OOTA, Y. and OSAWA, S. (1954) *Biochim. Biophys. Acta* **15**, 163.
OOTA, Y., YAMAMOTO, Y. and FUJII, R. (1954) *J. Biochem., Tokyo* **40**, 187.
OOTA, Y., FUJII, R. and SUNOBE, Y. (1956) *Physiol. Plant* **9**, 38.
OOTA, Y. and TAKATA, K. (1959) *Physiol. Plant* **12**, 518.
PAECH, K. (1935) *Planta* **24**, 78.
PEERS, F. G. (1953) *Biochem. J.* **53**, 102.
PRIANISHNIKOV, D. N. (1951) *Nitrogen in the Life of Plants*. Kramer Business Service.
REUHL, E. (1936) *Rec. Trav. Bot. Neerl.* **33**, 1.
ROSE, D. H. (1915) *Bot. Gaz.* **49**, 425.
ŠEBESTA, K. and ŠORM, F. (1956) *Coll. Czech. Chem. Comm.* **21**, 1047.

SEMENKO, G. I. (1957) *Fiziol. Rasteny* **4,** 332.

SMITH, B. P. and WILLIAMS, H. H. (1951) *Arch. Biochem. Biophys.* **31,** 366.

SPRAGG, S. P. and YEMM, E. W. (1959) *J. Exp. Bot.* **10,** 409.

STILES, W. (1935) *Bot. Rev.* **1,** 249.

STUMPF, P. K. (1952) *Ann. Rev. Plant Phys.* **3,** 27.

STUMPF, P. K. and BRADBEER, C. (1959) *Ann. Rev. Plant Phys.* **10.** 197.

SWITZER, C. M. and SMITH, F. G. (1957) *Canad. J. Bot.* **35,** 515.

TAZAKAWA, Y. and HIROKAWA, T. (1956). *J. Biochem., Tokyo* **43,** 785.

VIRTANEN, A. I., BERG, A. M. and KARI, S. (1953) *Acta Chem. Scand.* **7,** 1423.

WEBSTER, G. C. (1957) *Arch. Biochem. Biophys.* **68,** 403.

WEBSTER, G. C. (1959) *Nitrogen Metabolism in Plants.* Row. Peterson Biol. monographs.

WETTER, L. R. (1957) *J. Amer. Oil Chemists* **34,** 66.

YAMADA, M. (1955) *Sc. Paper Coll. Gen. Ed. Univer., Tokyo* **5,** 161.

YAMADA, M. (1955) *Sc. Papers Coll. Gen. Ed. Univer., Tokyo* **7,** 97.

YAMAMOTO, Y. (1955) *J. Biochem., Tokyo* **42,** 763.

YAMAMOTO, Y. and BEEVERS, H. (1960) *Plant Phys.* **35,** 102.

YOCUM, L. E. (1925) *J. Agron. Res.* **31,** 727.

YOUNG, J. L. and VARNER, J. E. (1959) *Arch. Biochem. Biophys.* **84,** 71.

YOUNG, J. L., HUANG, R. C., VENECHO, S., MARKS, J. D. and VARNER, J. E. (1960) *Plant Phys.* **35,** 288.

CHAPTER 6

THE EFFECT OF GERMINATION INHIBITORS AND STIMULATORS ON METABOLISM

WHENEVER a seed becomes dormant or ceases to be dormant the metabolism of the seed must change. Although this is an obvious conclusion to draw, in fact very little is known about the nature of the changes which occur. Experimentally it is inconvenient to have to wait till seeds become dormant or emerge from dormancy, as this does not permit an easy comparison of the different stages simultaneously. An alternative method of approaching the problem experimentally, is to induce dormancy or to break it. For this purpose some of the compounds known to affect dormancy, such as coumarin or thiourea, can be usefully employed. Comparatively little work along these lines has been carried out. Consequently, a great deal of the data in this chapter will draw upon the results of the authors, who used this method to study the metabolism of lettuce seeds.

In practice, the metabolism of the seeds under three different treatments was compared. Seeds germinated in water in the dark, having a germination of around 50 per cent, were taken as the control. These were compared with seeds whose germination was depressed to zero by the use of coumarin. As this inhibition was completely reversible by a number of treatments it was taken as representative of dormant seeds. Other seeds were treated with thiourea which induced 100 per cent germination. This treatment was therefore in some ways similar to natural dormancy breaking. In seeds so treated the metabolism of storage materials was investigated. A number of enzyme systems were studied, including especially overall respiration and the respiratory and electron-transport mechanism.

Dry lettuce seeds contain a very large amount of fat and a very small amount of free fatty acids. As normal germination

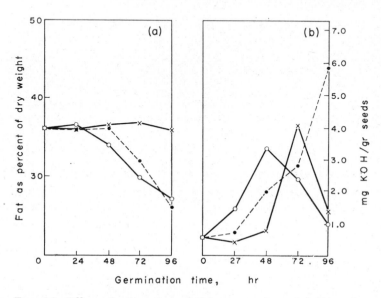

Fig. 6.1. Effect of coumarin and thiourea on fat metabolism of lettuce seeds during germination (Poljakoff-Mayber and Mayer, 1955).

(a) Changes in total lipids
(b) Changes in free fatty acids
O——O Seeds germinated in water
×——× Seeds germinated in coumarin (100 ppm).
●--- ● Seeds germinated in thiourea (1250 ppm).

proceeds, the lipid content of the seeds falls after about twenty-four hours. In the presence of coumarin this fall is prevented entirely, while in the presence of thiourea it is delayed for almost twenty-four hours [Fig. 6.1.(a)]. Free fatty acids, on the other hand, rise continuously in the seeds treated with thiourea, while in untreated seeds and seeds treated with coumarin the free fatty acid content first rises and then falls again [Fig.

6.1 (b)]. At least two lipases are apparently concerned with lipid metabolism, a neutral and an acid one. The acid lipase becomes totally inhibited by thiourea, while the neutral one increases markedly in activity. Coumarin on the other hand does not affect the acid lipase *in vitro* although it prevents its rise *in vivo*. The neutral lipase is somewhat inhibited by coumarin *in vivo* and considerably so *in vitro* (Fig. 6.2).

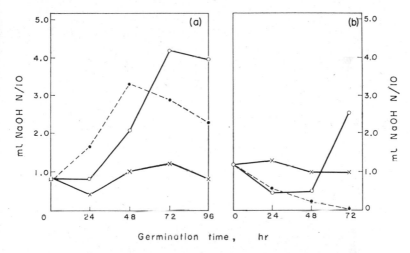

FIG. 6.2. Effect of coumarin and thiourea on lipase activity of lettuce seeds during germination (Rimon, 1957). Results as titer of alkali equivalent to fatty acids liberated by lipase.

(a) Neutral lipase; (b) Acid lipase
(Symbols as in Fig. 6.1)

These changes in lipid metabolism do not suggest any primary role for lipid metabolism in dormancy and dormancy breaking. In seeds germinated in coumarin the lipids are not metabolized, because the seeds do not germinate. In seeds germinated in thiourea some of the changes are accentuated but not altered fundamentally, during the initial step of germination.

The changes in lipid metabolism should be considered side by side with changes in sugars, into which the lipids may be converted. In Fig. 6.3 some of these changes are illustrated. It will be seen that during germination in water and in thiourea the metabolism of glucose and sucrose is essentially the same, although the rise of glucose in thiourea, is less than in water. In seeds germinated in coumarin, again, there is no glucose formation although sucrose is broken down.

Nitrogen metabolism in germinating lettuce seeds is characterized by a steady increase in the amount of soluble nitro-

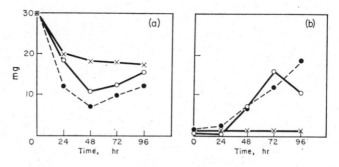

FIG. 6.3. Effect of coumarin and thiourea on sugar metabolism in lettuce seeds during germination (Poljakoff-Mayber, 1952). Results as milligram equivalents of glucose.
(a) Sucrose; (b) Reducing sugars
(Symbols as Fig. 6.1)

genous compounds, the total amount of nitrogen remaining constant for the first three days. This rise is directly proportional to germination percentage, the higher the germination percentage the greater the amount of soluble nitrogen. When germination is prevented in some way, e.g. by treatment with coumarin, this rise in soluble nitrogen is prevented (Table 6.1). This observation suggests that during normal germination, storage proteins are broken down and this breakdown is prevented by coumarin. This point of view is strengthened by the fact that coumarin can inhibit a proteinase present in the seeds (Poljakoff-Mayber, 1953). It is also supported by the fact that in seeds whose germination is promoted by light, the amount of soluble nitrogen increases more rapidly than in those germinated in darkness. Moreover, if seeds are placed

TABLE 6.1 — THE EFFECT OF COUMARIN ON CHANGES IN SOLUBLE NITROGEN
DURING GERMINATION OF LETTUCE SEEDS
Soluble N given as percentage of initial weight of seeds
(Klein, 1955)

Time of germination in hours	Seeds germinated			
	In water		In coumarin	
	% germ.	Soluble N	% germ.	Soluble N
0		0·260		0·260
24	17	0·228	0	0·264
48	46	0·365	6·5	0·272
72	50	0·556	11·5	0·273

under conditions suitable for germination and then, after various times, the germinated and non-germinated seeds are separated, the non-germinated ones show no increase whatever in soluble nitrogen while the germinated ones show a very marked increase (Klein, 1955). Such changes as have been observed in the storage materials all seem to be directly related to the germination process. In so far as coumarin depressed such changes, and light or thiourea stimulated them, this appears to be a direct result of a change in germination behaviour and not its cause. The same seemed to be true for the metabolism of such compounds as ascorbic acid and riboflavin in the seeds, as well as for the heavy metals which occur in various fractions of the seeds. In no case did germination inhibition or stimulation appear to effect these processes except in a secondary fashion, through their effect on germination.

Some marked effects of germination stimulators and inhibitors on the respiration of seeds have been noted. These observations relate both to the affect on the gas exchange of the seeds, and also to the biochemical mechanism of the respiration. Levari (1954) studied this effect on wheat and lettuce seeds not sensitive to light. The overall effect of coumarin on oxygen uptake and carbon dioxide output is by no means simple. At certain times coumarin raises the oxygen uptake somewhat above that of controls, but later the oxygen uptake is depressed. Another effect is the change in the respiratory quotient. Some of these changes, both for wheat and lettuce seeds, are illustrated in Fig. 6.4. A difficulty in interpreting

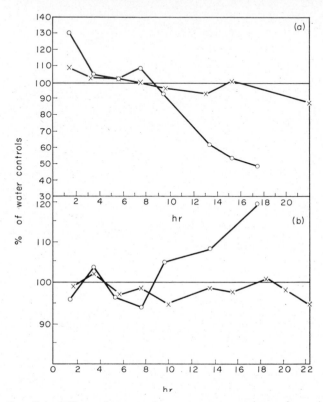

FIG. 6.4. Effect of coumarin on oxygen uptake and respiratory quotient of germinating seeds. (Levari, 1953). Results expressed as percent of water controls.

(a) Oxygen uptake;   (b) Respiratory quotient.

O——O Wheat seeds germinated in solutions of coumarin (250 ppm).
x——x Lettuce seeds, var. Progress (light-indifferent).
germinated in coumarin (100 ppm).

precisely this kind of result lies in the fact that coumarin in these experiments completely inhibited germination. It is fairly well established that respiration rises as germination proceeds. As a result, the comparison in Fig. 6.4 is between seeds whose respiration rises because they germinate, with seeds which do not germinate at all. Klein (1955) tried to overcome this difficulty by using light-sensitive lettuce seeds. These germinate relatively little in the dark. When treated with coumarin they do not germinate at all. However, by choosing a suitable light stimulus, the effect of coumarin can be overcome and the

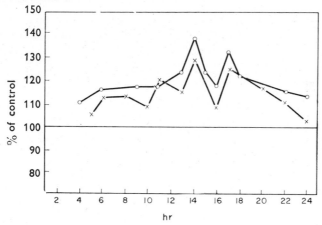

FIG. 6.5. Effect of coumarin on respiration of germinating lettuce seeds (Klein, 1955). The seeds of the light-sensitive variety Grand Rapids were germinated in 75 ppm coumarin and given a light stimulus two hours after beginning of imbibition. Germination was 7 per cent. Results as percent of water controls.

O——O Carbon dioxide output
×——× Oxygen uptake

germination restored to almost the same level as in the dark. Under these conditions the oxygen uptake of the coumarin-treated seeds is consistently higher than that of the water conrols (Fig. 6.5). The carbon dioxide output behaved similarly. However, it is not clear, in this type of experiment, whether the rise in respiration was due solely to the coumarin treatment or whether the light which increased germination also affected respiration. Quite similar results have been obtained by Ishikawa (1958) for pea and *Setaria* seeds. He again noted that coumarin at certain concentrations raises the oxygen uptake of the seeds. He also showed that this rise in oxygen uptake was quite similar to that induced in the same seeds by 2,4-dinitrophenol (DNP). Ishikawa suggests on this basis that coumarin is acting by uncoupling respiration from ATP formation in germination. This would be consistent with the fact that coumarin also increases the oxygen uptake of mitochondria isolated from lettuce seeds (Poljakoff-Mayber, 1955). Some of the more general observation on the effect of coumarin on metabolism during germination and growth (Mayer and Poljakoff-Mayber, 1961) could also be explained on the basis of uncoupling action although this is probably not the entire explanation of the observed facts. A further important feature of the course of respiration of lettuce seeds is the rise of respiration in two stages, previously mentioned, separated by a plateau. The first rise is not necessarily related to germination as such, and occurs equally in seeds which will eventually germinate and those which will not. The rise after the plateau occurs only in those seeds which germinate and is closely associated with seedling growth. If germination is prevented in some way, the second rise in respiration is usually also prevented. Even if germination is prevented by elevated temperatures, which increase the initial rise in the respiratory rate, the second rise in respiration is absent.

In contrast to these marked effects of coumarin on gas exchange, virtually no effect of thiourea on either Q_{O_2} or Q_{CO_2} was noted for lettuce seeds, nor did the RQ change appreciably as a result of germination in thiourea (Poljakoff-Mayber and Evenari, 1958).

Various attempts have been made to study the effect of coumarin on isolated enzyme systems involved in respiration. Glycolysis seems to be operating in peas whose germination was retarded or inhibited by coumarin and therefore the glycolytic system does not appear to be affected by coumarin.

Dehydrogenase activity of lettuce seed extracts is not affected by coumarin *in vitro*. However, the development of dehydrogenase activity *in vivo* is depressed appreciably by coumarin, presumably because the enzyme activity fails to develop (Table 6.2). Cytochrome-*c* oxidase activity is also not affected

TABLE 6.2 — THE EFFECT OF COUMARIN AND THIOUREA ON DIFFERENT
METABOLIC MECHANISMS IN GERMINATING LETTUCE SEEDS

Seeds germinated in / Metabolic change	Water	Coumarin 100 ppm	Thiourea 1250 ppm	Reference
Phytase activity of seeds germinated 48 hr. γ P released/g seeds	9600	2935	8880	Mayer, 1958
Phytin destroyed in seeds germinated for 48 hr. results as; γ P/g seeds	1510	120	1960	Mayer, 1958
as γ P/germinated seed	2·14	1·5	2·08	,, ,,
Malic dehydrogenase activity of seeds germinated for 48 hr γ Neotetrazolium reduced	60	20	—	Mayer *et al.* 1957
Phenolase activity of seeds germinated 48 hr as µl O_2/25 mg seeds/min Substrate: Quinol	10	9·5	0	Mayer, 1961
,, Methyl catechol	13·5	15·0	15·6	
Tricarboxylic acid cycle activity in seeds germinated for 14 hr as µl O_2/mg N/hr Substrate α-Ketoglutarate	5·0	5·0	13·0	Poljakoff-
,, Malate	5·0	5·0	8·0	Mayber and
,, Citrate	5·0	5·0	10·0	Evenari, 1958
Catalase activity in seeds germinated 48 hr as ml O_2/g seeds/min	600	500	120	Poljakoff-Mayber, 1956
Peroxidase activity of seeds as mg purpurogallin/g germinated 48 hr	7·5	5·0	39	Poljakoff-
germinated 96 hr	8·0	5·0	65	Mayber, 1953
DPNH oxidase activity of seeds germinated 48 hr Relative activity/mg N_2	100	93	0	Mayer, 1959

by coumarin *in vitro*. This suggests that coumarin does not act directly on these particular respiratory enzymes. Coumarin does however appear to effect phosphate metabolism. As already mentioned earlier, the main source of phosphorus in many seeds is phytin, which releases inorganic phosphate due to enzymic hydrolysis during germination. In lettuce seeds, phytin breakdown is completely prevented by germination in coumarin and the development of the enzyme phytase is considerably retarded. This is true even for those seeds which actually germinate in the solution of coumarin, whose phytase activity per seed was also lower than corresponding controls (Table 6.2). Coumarin may therefore interfere directly in the phosphate-releasing mechanism of these seeds. Moreover, in seeds germinated in coumarin, ATP failed to disappear, as it does during the first 24 hours in seeds germinated in water or thiourea. On the other hand hexose diphosphate was absent in seeds germinated in coumarin but present in seeds under the other treatments. Further, coumarin was shown to interfere directly in oxidative phosphorylation of mitochondrial preparations isolated from lettuce seeds, 130 ppm coumarin depressing phosphorylation by mitochondrial preparation from 1·5–2·5 micro-atoms/10 min to a value of 0·4–0·5 micro-atoms, when α-ketoglutarate was the substrate. The P/O ratio was depressed from 2·2 to about 0·3.

Thiourea did not affect phytin breakdown in any way. In contrast, enzymes of the tricarboxylic acid cycle were active at a much earlier stage of development when the seeds were germinated in a solution of thiourea than in non-treated controls (Table 6.2). This effect was very clear despite the fact that, *in vitro*, thiourea does not stimulate any of the enzymes of the tricarboxylic acid cycle and may even depress the oxygen uptake of isolated mitochondria in the presence of some of the substrates of the cycle. Oxidative phosphorylation also seems to be stimulated in seeds germinated in solutions of thiourea.

Certain other oxidative enzymes have been shown to be affected by coumarin and by thiourea. Catalase activity of lettuce seeds, germinated in solutions of thiourea, was strongly depressed almost immediately after the onset of treatment. Coumarin only very slightly depressed catalase activity. In contrast, peroxidase activity showed a steady and persistent rise in seeds germinated in solutions of thiourea but not in the controls. It is interesting to note that light, which also promotes germination of these seeds, had quite similar effects on both catalase and peroxidase (Table 6.2).

It has already been mentioned that other enzymes apart from cytochrome oxidase may function as terminal oxidases. Lettuce seeds contain an active phenolase which has a very high activity in dry seeds and which does not increase in amount as germination proceeds. The phenolase can also carry out the oxidation of ascorbic acid by coupled oxidation reactions. Although phenolase activity is only slightly depressed by thiourea, both *in vivo* and *in vitro*, the coupled oxidation of ascorbic acid is totally suppressed both *in vivo* and *in vitro* by thiourea. Other coupled oxidations carried out by this phenolase, e.g. of quinol, are affected by thiourea in the same manner (Table 6.2). A similar inhibitory effect of thiourea, both *in vivo* and *in vitro*, has been observed on a DPNH oxidase present in lettuce seeds. This latter enzyme may also be a phenolase. Coumarin did not have any effect whatever in any of these oxidation reactions.

A different aspect of the changes in metabolism caused by germination inhibitors and stimulators is their effect on the naturally-occurring, endogenous inhibitors and stimulators. In the first place, substances active in growth and germination, present in the seeds under different treatments, were directly extracted and assayed. Secondly, the interaction of externally-applied germination stimulators with other substances was investigated. Lettuce seeds were extracted so as to give two fractions, an acid and a neutral one. Each of these fractions, after chromatographic separation, was assayed in coleoptile growth, lettuce seedling root elongation and lettuce seed germination tests. In the acid fraction, after 12 hours of germination in solutions of coumarin, the formation of an additional coleoptile growth inhibitor was induced and the disappearance of the natural growth inhibitors initially present was prevented. Thiourea induced the formation both of an additional coleoptile growth promoter and an inhibitor, as well as the appearance of an additional germination promoter. In the neutral fraction the only marked effect was the appearance of a germination promoting substance, 2 hours after the seeds were placed in a solution of thiourea. These changes are therefore in agreement with what might be expected from the effect of these substances, i.e. coumarin causing, in general, the appearance of new inhibitors or the failure of ones already present to disappear, while thiourea, in contrast, induced the formation of new germination promoters (Blumenthal-Goldschmidt, 1958).

Interaction studies between coumarin and thiourea in their effect on germination and also between each of them and other substances were carried out. From a study of the interaction between coumarin and thiourea it appeared that coumarin, in essence, makes seeds more sensitive to thiourea, i.e. the maximum stimulatory effect on germination becomes evident at lower thiourea concentrations. Higher thiourea concentrations, which in the absence of coumarin caused a maximum effect, are much less effective in the presence of coumarin (Table 6.3). Coumarin and thiourea interact not only in their effect on germination but also in the subsequent growth of the seedlings. The similarity between thiourea and light in

TABLE 6·3 — THE EFFECT OF COUMARIN ON THE GERMINATION OF LETTUCE SEEDS IN THE PRESENCE OF THIOUREA
(Poljakoff-Mayber et al., 1958)

Thiourea concn. $\times 10^{-2}$ M	% germination			
	0	1·0	2·5	5·0
Coumarin concn. ppm				
0	33·0	57·8	59·1	63·4
10	2·7	15·6	34·3	14·4
20	0·8	8·5	13·8	3·6

stimulating germination is further reflected by the fact that coumarin makes seeds more sensitive to both agents (Mayer and Poljakoff-Mayber, 1961).

An additional interaction is that between coumarin and gibberellic acid. The latter can reverse the inhibitory action of coumarin in germination. This again points to similarities between the effect of gibberellic acid and light.

From studies on the uptake of thiourea by seeds, it appeared that thiourea stimulates germination while its internal concentration is still comparatively low, whereas the subsequent inhibition of seedling growth occurs when the internal concentration has risen appreciably. Thiourea shows interaction with a number of substances both in germination and growth. The kind of interaction usually differs for the two processes, possibly because of this concentration effect as well as because of the essential difference between germination and growth. The germination-stimulating effect of thiourea is closely linked to its structure and a change in its structure changes it from a stimulator to an inhibitor. When such a modified thiourea molecule is applied to seeds together with thiourea, each acts independently, the resulting germination being some kind of mean between stimulatory and inhibitory action. Despite the fact that thiourea can complex copper and despite its effect on copper-containing enzymes, its action cannot be simply attributed to its copper-complexing action, as the effect of thiourea on germination is not simply reversed by the addition of cupric ions. The relation between thiourea and copper ions in the seeds, as reflected in the activity of copper-containing enzymes, seems to be much more complex (see Table 6.2). Ascorbic acid, which itself does not affect germination at all, greatly enhances the stimulatory effect of thiourea. This effect was not due to interference in enzyme systems concerned with ascorbic acid metabolism but rather to some additional action of ascorbic acid (Poljakoff-Mayber and Mayer, 1961). From the results brought above it does not appear that any clear cut conclusion can yet be drawn about the mechanism by which coumarin inhibits germination or thiourea stimulates it. This is no doubt in part due to the lack of data on normal metabolism in seeds. Nevertheless a few possible modes of action seem to emerge.

Coumarin is often supposed to function by blocking SH groups. This is primarily based on the reversal of coumarin inhibition in growth by BAL. However Ishikawa (1958), showed that BAL does not ordinarily reverse germination inhibition caused by coumarin, and Mayer and Evenari (1952) on the basis of structure-activity studies, concluded that a blocking of SH groups could not explain coumarin action.

Coumarin acts by affecting respiratory metabolism both directly, by interfering at the phosphorylation stage, and indirectly, by affecting the availability of phosphate. It may act by preventing the formation of certain enzyme systems possibly because it prevents their liberation from some bound form. Finally, it may interfere with the formation and destruction of endogenous substances regulating germination.

Thiourea also acts by affecting the respiratory mechanism, possibly by rapidly channelling all respiration in the direction of energy yielding processes. A further way by which thiourea may act is by changing the nature and amount of the germination regulators present in the seeds.

The effect of coumarin and thiourea on metabolism of germinating seeds has been dealt with at some length because these substances have rather specific effects in seeds during germination. Such investigations serve to indicate some of the problems and achievements in metabolic studies during germination even though they do not solve them unequivocally.

Other substances which have more general effects on metabolism can inhibit germination. The effect of these substances on the metabolism of germinating seeds has been studied to some extent in order to determine whether their effect here is the same as in metabolism in general. Special attention has been paid to the effect of 2,4-dinitrophenol (DNP). The effect of DNP on the germination of pea and lettuce seeds is shown in Fig. 6.6. Lettuce seeds are apparently much more sensitive to DNP than are peas if the concentration giving 50 per cent inhibition is considered. The inhibition of germination by DNP, in lettuce, was not reversible by light. DNP up to concentrations of 10^{-3} M did not effect either

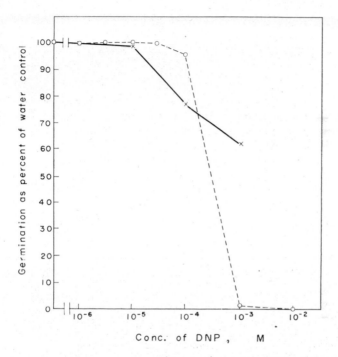

FIG. 6.6. Effect of various concentrations of dinitrophenol (DNP) on germination of lettuce and pea seeds. Results as percent of germination of the water controls (compiled from the results of Poljakoff-Mayber, unpublished; Ishikawa, 1958).

O——O Lettuce
×——× Peas

glucose or sucrose metabolism in lettuce in any way. The metabolism of nitrogenous compounds was somewhat affected by inhibitory concentrations of DNP, but the effects were not clear-cut, and did not account for inhibition of germination. At low concentrations DNP raises the oxygen uptake of seeds in a fashion entirely similar to that observed in other tissues. In lettuce, oxygen uptake was increased by about 30 per cent by a DNP concentration of 5×10^{-4}M and the carbon dioxide output was raised about 50 per cent above the controls (Klein, 1955). In peas, oxygen uptake was increased about 10–15 per cent by a DNP concentration of 10^{-4} M (Ishikawa, 1958). In both cases these concentrations coincided reasonably

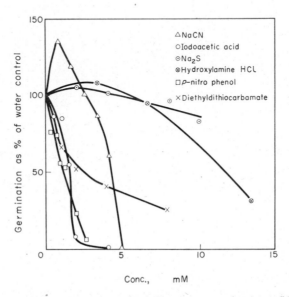

FIG. 6.7. Effect of respiratory inhibitors on germination of lettuce seeds (Mayer, Poljakoff-Mayber and Appleman, 1957). The seeds used were of the light-sensitive variety Grand Rapids. Tests carried out at 26°C in the dark.

well with those which inhibited germination. It seems, therefore, entirely reasonable to assume that DNP inhibits germination because it acts as an uncoupling agent which interferes with the energy supply of the germinating seeds.

Other respiratory inhibitors which have been studied are cyanide, azide, hydroxylamine, ethionine, diethyldithiocarbamate, iodoacetate and p-nitrophenol. The effect of some of these substances on germination are shown in Fig. 6.7. Ethionine at 5×10^{-3} M and azide at 10^{-3} M also inhibited the germination of lettuce. The effect of ethionine was partially reserved by light but that of azide was not (Mancinelli, 1958).

When the effect of some of these substances on the respiration of seeds was studied, it was found that the response differed when whole seeds were treated from that obtained when homogenates were used. There was some indication that the permeability of whole seeds was a factor in the response obtained. Different inhibitors penetrated the seeds at different rates, some very quickly. Among those which penetrated rapidly were cyanide and diethyldithiocarbamate. When the inhibitors were added to homogenates prepared from seeds germinated in water the effects were observed immediately. However the effect obtained differed, depending on the period for which the seeds had been germinated (Table 6.4). Ethionine

TABLE 6.4 — RELATIVE OXYGEN UPTAKE OF WHOLE SEEDS AND SEED HOMOGENATES IN THE PRESENCE OF VARIOUS RESPIRATION INHIBITORS ADDED EITHER *in vivo* OR *in vitro*. (The oxygen uptake of water germinated seeds or homogenates prepared from them is taken as 100. Oxygen uptake measured after 60 min.)
(Mayer, Poljakoff-Mayber and Appleman, 1957)

Inhibitor added during germination or to Warburg vessels	Concn. mM	Oxygen uptake of whole seeds		Oxygen uptake of homogenates of seeds	
		Germinated in water, inhibitor added to Warburg vessel	Germinated in inhibitor	Germinated in water, and inhibitor added to homogenate	
		18 hour	18 hour	18 hour	48 hour
(1)	(2)	(3)	(4)	(5)	(6)
Sodium cyanide	0·8	82	125	47	41
Diethyldithio-carbamate	1·0	86	108	127	114
Hydroxylamine-hydrochloride	10·0	114	66	0	26
Iodoacetate	2·0	105	22	90	95
Sodium sulphide	10·0	100	70	60	58
p-nitrophenol	1·0	100	87	105	76

and azide, at concentrations which inhibit germination, depressed oxygen uptake by 20–50 per cent and also changed the RQ of the seeds (Mancinelli, 1958).

Again it appears that the germination-inhibiting effect of these compounds is consistent with their effect on respiration, which is more general and applies to many other tissues. In so far as the response in seeds and seedlings differed, this can be ascribed to the fact that not all the respiratory mechanisms are functioning equally at different stages of germination. Moreover, it appears that at least some of these compounds can be broken down by the seeds, for example diethyldithiocarbamate.

Penicillin and streptomycin can also inhibit germination (Mancinelli 1958) and respiration was inhibited at the same time (Table 6.5). However, at low concentrations these substances could stimulate germination, an effect which was apparently reversed both by light and by higher temperatures. From interaction studies with light and temperature Mancinelli concluded that the germination-inhibitory effect of streptomycin could be accounted for entirely by its inhibition of respiration, but that this was not true for the action of penicillin. However,

TABLE 6.5 — EFFECT OF SOME METABOLIC INHIBITORS ON RESPIRATION
OF LETTUCE SEEDS
Respiration as µl gas exchanged. Seeds germinated for 16 hours at 22°C
(Compiled from Mancinelli, 1958)

	Water	Ethionine 5×10^{-2}M	Sodium azide 10^{-3}M	Penicillin $3 \cdot 5 \times 10^{-3}$ M	Streptomycin $8 \cdot 5 \times 10^{-4}$ M
Q_{O_2} (µl)	66	43	32	37	31
Q_{O_2} % control	100	65	48	56	47
Q_{CO_2} (µl)	56	34	39	28	24
Q_{CO_2} % control	100	60	62	58	43
RQ	0·85	0·79	1·10	0·76	0·77
RQ % control	100	93	129	89	90

no further metabolic studies were made so that these conclusions must at present be regarded, at best, as tentative.

An entirely different type of compound affecting germination is that of the herbicides. These do not in fact present any homogeneous group since many widely different substances are used as herbicides. Many of these substances when applied directly to seeds will prevent their germination. In agricultural practice herbicides are not normally used to prevent germination. They are, however, frequently used to kill the freshly-emerging weed seedling, while leaving untouched or less affected the deeper sown crop seeds. In this case their selectivity is chiefly based on depth of penetration into the soil, rather than on any inherent physiological differences between the species to be eradicated and that to be left. Very little is known regarding the effect of herbicides on the metabolism of seeds. Their effect on metabolism in general has also not been studied very extensively, except for a few cases. The compound most widely investigated has been 2,4-dichlorophenoxy acetic acid (2,4-D). This substance has been taken to be representative of all auxin-type herbicides and has served as a model in many reactions in which the mode of action of indolyl acetic acid was being investigated. Generally speaking, 2,4-D induces profound changes in the metabolism of treated plants. These occur in the metabolism of nitrogenous compounds, carbohydrates and to some extent respiration. Thus 2,4-D has been found to cause a more rapid use of carbohydrates, to interfere in the normal regulating action of endogenous growth substances, possibly in some case by increasing the amount of unsaturated lactones of the coumarin type such as scopoletin and methyl umbelliferone in the tissue, and finally to change the ratio between soluble and non-soluble nitrogenous compounds, by increasing the former. Scattered evidence points to 2,4-D increasing inorganic phosphate, due to decreases in the amount of phosphorylated sugars. Many enzymes have been shown to be affected to a greater or lesser extent by 2,4-D. These effects have not, however, been such as to account for the herbicidal action of

2,4-D. In fact, so far no single unambiguous effect of 2,4-D on metabolism has been observed.

In germinating lettuce and wheat no changes in carbohydrate metabolism were induced by 2,4-D during the first 24 hours of germination (Levari, 1960). It is of course possible that at a later stage effects similar to those noted in other tissues would result.

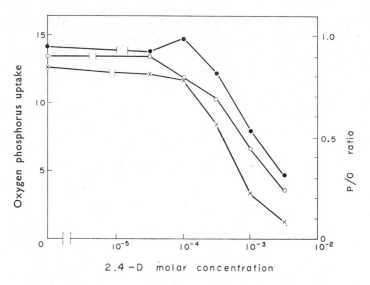

Fig. 6.8. Effect of 2,4-D on oxygen uptake and phosphorylation by soybean mitochondria, with succinate as substrate (compiled from data of Switzer, 1957).

     O——O   Oxygen uptake.
     ×——×   Phosphorus uptake.
     ●——●   P/O ratio.

Respiration of wheat and lettuce was increased somewhat by 2,4-D, and this increase was accompanied by changes in the respiratory quotient. Further evidence for the effect of 2,4-D on respiration comes from experiments with isolated mito - chondria. In these, increasing 2,4-D concentrations progres - sively inhibited oxygen uptake and phosphate uptake and lowered the P/O ratio (Fig. 6.8). A further interesting effect of 2,4-D was found in the carbohydrate metabolism of root tips of seedlings of a number of plants. In such root tips, 2,4-D when applied to intact seedlings, depressed glycolysis and increased metabolism via the pentose cycle. If applied to seedlings from which the cotyledons had been removed, glycolysis was not depressed but metabolism through the pentose cycle was nevertheless increased (Humphreys and Dugger, 1957). Lately this has been further supported by the findings that 2,4-D treatment actually increases the amount of enzymes which participate in the pentose cycle (Black and Humphreys, 1960).

From the above discussion it appears that many changes in metabolism are evoked by substances which *inter alia* also inhibit germination. Of the substances discussed, only for DNP and possibly for streptomycin, can their germination-inhibiting action be directly ascribed to changes in phosphorylation and respiration respectively. In all the other cases there is no possibility of relating their effects on germination to any definite metabolic process. There is no reason at present to assume that germination can be inhibited by only one kind of mechanism. It is not evident even that any given germination inhibitor acts on a single metabolic process. On the contrary, it appears that many compounds which inhibit germination do so simply because they interfere in a general way in normal metabolism. It is, however, possible that specific germination inhibitors act on some single stage of metabolism. Too little data are at present available to reach any definite conclusion on this point.

BIBLIOGRAPHY

BLACK, C. C. and HUMPHREYS, T. E. (1960) *Plant Phys.* 35, XXVII.
BLUMENTHAL-GOLDSCHMIDT, S. (1958) Ph.D. Thesis. Jerusalem (in Hebrew).
HUMPHREYS, T. E. and DUGGER, W. M. (1957) *Plant Phys.* 32, 136.
ISHIKAWA, S. (1958) *Kummamoto, J. Science Ser.* B 4, 9.
KLEIN, S. (1955) Ph.D. Thesis, Jerusalem (in Hebrew).
LEVARI, R. (1953) *Pal. Bot. Jer. Ser.* 6, 47.
LEVARI, R. (1960) Ph.D. Thesis, Jerusalem (in Hebrew).

MANCINELLI, A. (1958) *Ann. di Botanica* **26,** 56.

MANCINELLI, A. (1958) *Ann. di Botanica* **26,** 67.

MAYER, A. M. (1958) *Enzymologia* **19,** 1.

MAYER, A. M. (1959) *Enzymologia* **20,** 313.

MAYER, A. M. (1961) *Physiol. Plant* **14,** 322.

MAYER, A. M. and EVENARI, M. (1952) *J. Exp. Bot.* **3,** 246.

MAYER, A. M., POLJAKOFF-MAYBER, A. and APPLEMAN, W. (1957) *Phys. Plant* **10,** 1.

MAYER, A. M. and POLJAKOFF-MAYBER, A. (1961) in *Plant Growth Regulation,* Iowa State University Press

POLJAKOFF-MAYBER, A. (1952) *Pal. J. Bot. Jer. Ser.* **5,** 186.

POLJAKOFF-MAYBER, A. (1952) *Bull. Res. Council, Israel* **2,** 239.

POLJAKOFF-MAYBER, A. (1953) *Enzymologia* **16,** 122.

POLJAKOFF-MAYBER, A. (1953) *Pal. J. Bot. Jer. Ser.* **6,** 101.

POLJAKOFF-MAYBER, A. (1955) *J. Exp. Bot.* **6,** 313.

POLJAKOFF-MAYBER, A. and MAYER, A. M. (1955) *J. Expt. Bot.* **6,** 287.

POLJAKOFF-MAYBER, A. and EVENARI, M. (1958) *Physiol. Plant* **11,** 84.

POLJAKOFF-MAYBER, A., MAYER, A. M. and ZACK, S. (1958) *Bull. Res. Council, Israel* **6 D,** 117.

POLJAKOFF-MAYBER, A. and MAYER, A. M. (1961) *Ind. J. Plant Phys.* **3,** 125.

RIMON, D. (1957) *Bull. Res. Council, Israel* **6 D,** 53.

SWITZER, C. M. (1957) *Plant Phys.* **32,** 42.

CHAPTER 7

THE ECOLOGY OF GERMINATION

In previous chapters the effect of various factors on germination of seeds has been considered. In the following an attempt will be made to relate these observations to the behaviour of seeds in their natural habitat. Various mechanisms regulate the germination of seeds, some of which are internal, whereas others are external environmental factors, any of which can determine whether a given seed will germinate in a certain place or not. However, the demonstration in the laboratory of the existence of a regulating mechanism is not proof of its operation under natural conditions. To prove this and to show that the plant derives some definite advantage from a regulatory mechanism is much more difficult. The only advantage about which one may justifiably speak is where some special mode of germination has survival value for the species. This implies that the existence of the mechanism under study enables a species to exist under a given set of conditions, while the absence of this mechanism will prevent it from surviving. Survival value may also take the form of giving a species a better chance to establish itself in competition with other species occupying the same habitat. Unfortunately the proof of survival value of germination-regulating mechanisms is not easily obtainable and relatively few detailed studies have been made.

A spread of germination over a period of time can have survival value. This can protect the species from eradication as a result of adverse conditions which follow germination. If all seeds germinated simultaneously, then no further renewal of the species would be possible.

The ecological conditions prevailing in a given habitat will affect germination. In this respect probably not overall climatic conditions but rather the micro-climatic conditions prevailing in the immediate vicinity of the seed will be the determining factors. Normally speaking, seeds are shed so as to fall either on soil or leaf litter, or in some cases they may fall in a region more or less covered with water. In other words, seeds are usually situated in or on the soil or in water. The conditions prevailing under these circumstances will depend on the nature of the soil, its chemical composition and its physico-chemical structure and on their depth in the soil or under water. Depth will influence aeration as well as penetration of light. The chemical composition of the soil, or of the water, may affect germination in a number of ways. Possibly the soil may consist of leaf litter or partially decomposed matter which contains substances inhibiting germination. The soil may have a salt content which will osmotically retard or prevent germination or it may have an ionic balance unfavourable to germination. Soil structure will, in addition to affecting aeration and water content, also determine the ability of a seedling to emerge above the soil and establish itself or, in the case of a seed falling on the soil, the ability of the roots of the seedling to establish themselves in the soil. We will first try to enumerate the various external factors affecting germination and their relation to various types of habitat. Following this, we will try to relate the existence of certain germination-regulating mechanisms in seeds to factors in the habitat which might be connected with such mechanisms.

I. External Factors as They Appear in Various Habitats

1. *Water*

The water content of soils can vary from saturation, as in swamps and waterlogged soils, to zero or near zero for sandy soils in arid regions. Again, the precise water content of a given soil can vary with climatic conditions, coverage of the soil by plants and especially seasonally, with rainfall. The moisture content of the soil may vary within wide limits, not only between different soil types, but also in the same soil at different times of the year. Frequently habitats are classified according to their total moisture content, as hydric, mesic or xeric. However, for germination, the availability of water at a given period of time is the determining factor. Availability will be determined by osmotic factors, binding of water by soil colloids, capillary forces and soil composition and texture. In addition, of course, availability of water will be determined by competition with other organisms requiring and competing for water.

The moisture content of the soil may show seasonal periodicity. In some regions, high moisture content is associated with high temperatures, i.e. when summer rains are followed by dry winters or by very cold winters with frozen soil, and in other regions with low temperatures, i.e. when winter rains are followed by hot, dry summers. Sometimes, high moisture content occurs when the soil temperatures are at or near freezing so that, although moisture is present, it is in fact not available. Seasonal variability in moisture content is, however, by no means universal, e.g. in those regions where precipitation is more or less equally distributed throughout the year. On the other hand, where rainfall is seasonal, where summer rains or winter rains occur, the moisture content of the soil can vary from very high, immediately after heavy precipitation, to near dryness before the next rain, even within the rainy period.

A special condition exists in habitats of high salinity. Soils may be coastal, where the source of salinity is chiefly sea spray which, by causing accumulation of salt in the soil, also changes water availability. Inland saline soils are frequently marked not only by high salt content but also by being relatively impervious, having poor drainage and by being often, at any rate for part of the year, flooded. Furthermore, certain areas are characterized by periodic flooding, the plants growing in the periods between successive floodings. A very special case is constituted by the mangroves which actually grow with their root-systems submerged in sea water. On the other hand certain regions, such as some areas around the Dead Sea, have a very high salt content but are practically dry throughout the year because of absence of rain and relatively high temperatures all the year round.

2. Temperature

The soil temperature is extremely variable and shows both diurnal and seasonal changes. The extent of the changes depends on the one hand on the type of the soil (heavy or light, with or without leaf litter) and on the other hand, on the climatic conditions prevailing. In addition, very steep gradients of temperature with depth usually exist. These again will depend on the type of the soil. Soil texture and structure as well as the amount of water present in the soil, conditions for evaporation of water from the soil, and plant cover, all play a part in determining soil temperature. Usually, the upper layers of the soil show wide fluctuations and as greater depths are reached, conditions become more and more constant throughout the year.

3. Gases

The composition of the gaseous phase in the soil can be variable. Oxygen, nitrogen and carbon dioxide are the three gases normally present. As the equilibrium between the air above and the gaseous phase in the soil is attained only very slowly, appreciable differences between the phases may exist. The lighter the soil, the lower its organic matter content and the smaller the number of micro-organisms in it, the greater will be the correspondence between composition of the air in the soil to that above it. In soils having a high organic content and containing many micro-organisms the carbon dioxide content may be very much higher and the oxygen content much lower than in the air. In waterlogged soils and especially in heavy soils the oxygen content of the gaseous phase may drop considerably below that normal in the atmosphere. This is also true, in general, for soils having appreciable vegetation. In such soils the roots of the plants will take up oxygen and produce carbon dioxide, again changing the balance of gases.

The volume of the gaseous phase differs greatly in diffe- rent soils, being greatest in non-compact soils and least in heavy, compacted soils or waterlogged soils. In addition to the three main gases, the soil may contain others, chiefly due to the activity of micro-organisms and the absence of oxygen. Thus, soils may contain methane, hydrogen sulphide, hydrogen, nitrous oxide and probably also small amounts of carbon monoxide and ammonia.

4. *Light*

Light is abundant usually only on the surface of the soils. In light or sandy soils light penetrates a short distance into the soil, although its intensity falls off rapidly. In heavy soils light hardly penetrates at all. In cases where the soil is covered by water, light will penetrate considerable distances provided the water is clear.

In germination any two or all of the factors mentioned can show marked interaction. We will now attempt to see what part these various external factors can play in regulating ger- mination.

II. Ecological Role of External Factors

1. *Moisture and Temperature*

Probably the most crucial factor in determining germina- tion of seeds in the soil is a suitable combination of tempera- ture and moisture. As seeds are a means of propagating the species, their germination should occur at a time which will favour survival of the seedling. This time very often does not coincide with the time of shedding of the seeds. Despite the fact that moisture is often present when the seeds are shed, the seeds may not germinate. This may be because they have some special temperature requirement which postpones germination for some time after the seeds have been shed. The value of such a temperature requirement is obvious in the case of plants which shed their seeds in autumn. Immediate germination would lead to the killing of the seedlings in the subsequent cold winter. Postponement of germination by some sort of dormancy, which may be due to a temperature requirement or some other factor, will give the developing seedling a better chance for survival. Went (1957) cites examples of this type of behaviour for a number of plants growing in the Colorado desert. When soils from this desert were moistened in the laboratory at low temperature (10°C), chiefly winter annuals germinated. If they were moistened at higher temperatures (26–30°C) only summer annuals germinated, while at intermediate temperatures yet a third group of plants could be recognized. These groups corresponded very well to those actually occurring in the natural habitat. The flora resulting from summer rains included such plants as *Amaranthus fimbriatus*, *Euphorbia micromera*, *E. setiloba* and *Portulaca oleracea*. The flora after winter rains included various species of *Gillia*, a variety of small *Oenothera* species, including *O. palmeri*, and *Mimulus bigelovii*. The third group which occurred at intermediate germination temperatures in the laboratory, corresponded in nature to plants germinating after rains falling in very late autumn or early winter. Winter rains occur very regularly each year while the summer rains are far more infrequent. Apparently, therefore, the summer annuals must have a fairly extended viability as a condition for survival, a condition which is not an obvious essential for the winter annuals. Some seeds in this desert will only germinate very shortly after ripening, e.g. *Yucca brevifolia*. Apparently in this plant, if the seeds fail to germinate, they are rapidly destroyed by attacks by rodents and insect larvae. Usually only very few seem to germinate at all, and then only if summer rains follow immediately after the seeds are shed. These examples show the significance of the combination of temperature and moisture in determining germination and survival of the species. A requirement for a fixed temperature is not an invariable rule. Often alternating temperatures are required which constitute an adaption to certain climatic conditions. Juhren *et al.* (1953) investigated the germination of various grasses and especially of a number of species of *Poa*. They found that although all the species they studied could germinate and develop at moderate temperatures and with moderate diurnal alternations of temperatures, only one species, *Poa pratensis*, could stand diurnal changes of 26°C by day, 20°C by night in association with summer sunlight. This condition is one which the plant in question would occasionally meet in its normal habitat. On the other hand it did not germinate at all in the cold. *Poa scrabrella* and *P. bulbosa* apparently were adapted to germination in the cold, day and night 3·5°C, or day 6°C and night 3·5°C. These plants occur in regions where cold conditions do prevail. *Avena fatua* seems to germinate particularly well after frequent alternations between freezing and thawing, followed by periods of wetting and drying. This treatment corresponds to the climate in those periods in the year in Bavaria when massive infestations by this weed occur (Bachthaler, 1957).

In a number of plants germination only occurs if the seeds are shed in a very moist habitat. An example of such seeds is willow, which is characterized by a very rapid loss of viability. The presence of the trees in very moist habitats may therefore be related to this property of the seeds. Another species, the seed of which only germinates under extremely moist conditions, under water, is *Taxodium distichum*. The requirement seems, however, to be more for an anaerobic condition rather than for the very high moisture content of the soil. This species of *Taxodium* grows characteristically in swamps.

Seeds of many plants will not normally germinate under water and frequently, if such seeds are kept under water for any length of time, their viability becomes impaired. Other species can withstand being placed under water and will even germinate under such conditions. Frequently, however, germination is greatly improved by increased aeration. The moisture tension under which different plant seeds will germinate can be very different. Some species will germinate under a wide range of conditions of moisture tension, while others are much more limited in their requirements. The number of seeds which germinate under water is comparatively small. It includes such species as *Roripa nasturtium* (water cress) and *Typha latifolia* (Morinaga, 1926). Again this requirement seems to be not under water conditions as such, or for the presence of an excess of liquid water, but a requirement for the anaerobic conditions which are associated with this. The case of *Typha* has been considered in some detail by Sifton (1959). This plant will normally germinate under water if given a light stimulus. The seeds will germinate not only under water but also in normal conditions, e.g. on moist filter paper, provided the oxygen tension is lowered (Table 7.1). Sifton found that germination is not simply conditioned by light and oxygen tension but also by temperature. The general conclusion reached from this study was that the germination of *Typha* was determined by the amount of water which can be held by the colloidal protein of the aleurone grain. The water-holding capacity of these grains is increased by illumination with white light and by products of anaerobic processes.

TABLE 7.1 — GERMINATION OF *Typha latifolia* L. SEEDS UNDER VARYING CONDITIONS
(Compiled from data of Sifton, 1959)

	Seeds immersed in water	% germination	
		Seeds on moist blotting paper	
		In air	In 2% O_2
35°C Light	81	48	—
30°C Light	89	61	96
30°C Dark	4	—	—
25°C Light	86	44	—
20°C Light	69	37	—
15°C Light	24	20	—

However, if respiration is very vigorous, vacuolation in some of the cells is said to occur and the resulting turgor pressure can compensate for the lack of swelling of the aleurone grains, thus again permitting germination. This would account for the observation that some of the seeds will germinate even in air.

An adaptation of germination and seedling survival under anaerobic conditions also seems to exist in rice (see Chapter 3), but in rice, at any rate, this is not an obligatory habitat or even one desirable for seedling development, while in *Typha* anaerobic conditions were the preferred ones.

Instances are known where submergence of seeds in water, followed by exposure to air, promotes germination. Seeds of both *Heliotropium supinum* and *Mallugo hirta* will germinate only if they are buried in wet mud for a certain period of time and then exposed to air. These seeds have an additional requirement for germination, viz. low temperature for a period of time (Mall, 1954). It appears that continuous soaking increases the permeability of the testa. Both these plants occur in the drying pools and puddles which form following monsoon rains. Thus it might be possible that the conditions required constitute an adaptation to a very specific habitat. However, another plant, *Polygonum plebejum*, seems to occur in the same habitat and yet does not require pre-soaking for its germination.

An entirely different problem exists in plants growing in arid regions. In these plants, survival of the species is determined by mechanisms which ensure that germination occurs at a time when the seedling will be able to establish itself. Thus, ideally, germination should occur when moisture and temperature conditions favour both germination and seedling growth. A variety of mechanisms has at different times been suggested as operating in the seeds of various plants growing in such conditions. However, some advantage might result from the mere prevention of germination under exceptionally unfavourable conditions. A survey of the known germination behaviour of desert seeds does not give any very clear picture, but germination inhibitors seem to play a role. Some of the earliest experiments were those of Went. He suggested that in certain desert seeds germination only occurs if the rate of reformation of an inhibitor in the seeds, when they are moistened, is slower than the process of germination itself. Such a situation might occur only under very specific conditions of moisture and temperature (Went, 1953 and 1957). Thus in seeds of *Pectis pappoza* germination occurs only after 25mm of summer rains and not after an equal or greater amount of rain in the winter. A different regulatory mechanism connected with rainfall has been described by Soriano (1953) for *Baeria chrysostoma.* Seeds of this plant germinate only if they are sown on wet soil. However, the subsequent development of the seedling was found to be dependent on the amount of additional precipitation which the seeds received. In the absence of additional precipitation, total germination was lower and seedling development poorer, and fewer flowers were formed than in the presence of an added 50 mm of rain immediately after sowing on wet soil.

Went, on the basis of these and many other instances of the behaviour of desert seeds, believes that the seeds contain one or more inhibitors which separately or together act as a kind of rain gauge which determines when germination will occur. It is supposed that the inhibitor content at which germination occurs is correlated to the amount of rainfall which will permit seedling establishment. Evidence for the presence of inhibitory substances in seeds or fruits of certain desert plants is available (Koller, 1955; Koller and Negbi, 1959). However, it is unfortunate that the nature of these inhibitors has never been ascertained. Moreover, quantitative changes in them have never been proved experimentally. Thus, attractive as this theory is, it is still lacking final experimental proof. Undoubtedly this mechanism is not universal and others also occur. Thus Koller and Cohen (1959) showed that seeds of three *Convolvulus* species occurring in arid areas are impermeable to water. The seeds germinate only if their permeability is raised mechanically by abrasion or impaction, or chemically, by acid treatment. In addition, a fairly high temperature is also required. Lack of permeability prevents germination. However, the seeds become permeable gradually over a long period of time. It is, therefore, possible that each year there will be a certain number of seeds which can imbibe water and subsequently germinate, when temperature conditions become favourable. In this way not all the seeds will germinate together. If, subsequent to germination, the seedlings are destroyed, the entire population will not be wiped out, as part of the seeds will still be impermeable and viable. Survival of the species may be ensured even if germination is only restricted by lack of permeability to water and is not regulated as such. Impermeable seed coats are of very frequent occurrence in many seeds, especially among the Leguminosae. This characteristic of the seed coat does not appear to be restricted to plants occurring in dry habitats. The germination of *Calligonum comosum*, a desert shrub growing in the Sahara and similar habitats in nature appears to be limited to sandy environment. The germination behaviour in this case seems to restrict the plant to such habitats as coarse sandy soils with low rainfall. This appears to be due to the fact that the seeds will not germinate if in contact with liquid water for any length of time. Again the germination behaviour is complicated by light sensitivity of the seeds, which germinate better in the dark than in the light. Under natural conditions germination seems to occur only if the seeds are buried in sand (Koller, 1956).

Another type of adaptation to moisture is met with in the seeds of *Panicum antidotale*, *P. turgidum* and *Artiplex dimorphostegia* (Koller, 1954). Seeds of these plants germinate to a higher percentage in the laboratory if the seeds, together with their dispersal unit, are dried over calcium chloride before sowing. In their natural habitat the seeds are often exposed to hot dry weather particularly in the direct sun, before they germinate. The mechanism of pre-drying might thus constitute a special adaptation to their normal habitat. It is, however, not clear what is the precise relative humidity in the micro-climate surrounding the seeds. Pre-drying is certainly not an

essential for germination, as part of the seeds germinate even
without it and the whole mechanism only points to a correlation between habitat and germination behaviour.

2. *Gases*

From an ecological point of view, the function of gases in
regulating germination is very puzzling. The normal requirement is for aerobic conditions, i.e. 20 per cent oxygen in the
atmosphere. As discussed previously, germination of many
seeds is prevented when the oxygen content is appreciably
lowered. However, the seeds of certain plants and particularly
of aquatics such as *Typha* prefer anaerobic conditions for germination. Morinaga, when testing a variety of seeds, found
that among 70 species at least 43 were capable of germinating
under water, i.e. under anaerobic conditions 18 out of these;
43 species germinated equally in water and in air on moist
filter paper. Only two species had a definite preference for
anaerobic conditions. The only natural habitats where one would
expect to meet anaerobic conditions are under water, and in
waterlogged soils and swamps. Nevertheless, in all kinds of
habitat the composition of the atmosphere surrounding seeds
can change. The ratio of oxygen to carbon dioxide in particular is liable to change, for example, as a result of the activities of the micro-flora and fauna in the soil. This may lead
to an increase in the carbon dioxide concentration with a simultaneous decrease in the oxygen content of the atmosphere in
the soil. A different way in which the gaseous environment of
the seeds may change arises from the differential permeability
of the seed membranes to different gases. This can lead to an
increase of carbon dioxide and a decrease of oxygen within
the seed. Both in the soil and in the seed the changes in the
gaseous atmosphere are confined to the oxygen and carbon
dioxide fractions, as the nitrogen content usually remains
constant.

No precise information on the internal atmosphere of the
seed is available. The only attempt to measure this seems to

have been made by Kidd (1914) in peas (Table 7.2). Kidd ground peas either in water or in barium hydroxide. Those ground with water were allowed to equilibrate with the air and then barium hydroxide was added. The difference in titer between the two treatments was taken as a measure of the carbon dioxide content of the seeds. This method must be regarded with considerable reserve as regards the carbon dioxide content of the seeds, and does not indicate the carbon

TABLE 7.2 — THE CO_2 CONTENT OF
PEAS, *Pisum sativum*, WHILE
GERMINATING
(after Kidd, 1914)

Time of germi-nation (hr)	CO_2 content mlCO_2/100 g of seeds
Dry Seeds	145
18	64
25	41
39	43
64	39
97	16

dioxide/oxygen ratio within the seed. All other conclusions on the gaseous atmosphere in the seeds are based on the response of seeds, with or without seed coat, to changes in the composition in the external atmosphere. This need not always correspond to a change in the internal atmosphere.

Some information is available about the permeability of isolated membranes from seeds, as described earlier (Chapter 4). Kidd concluded from his experiments that internal carbon dioxide accumulation, above a certain level, inhibits germination. Germination will occur only when this level falls. According to Kidd such a drop occurs in peas about 18 hours after the seeds are placed in water (Table 7.2). If, in fact, this conclusion is correct, then this may possibly regulate germination. As long as internal carbon dioxide concentrations are high, germination is retarded, even if the seed is imbibed.

This may even induce secondary dormancy. When the impermeable membrane is destroyed or punctured, the accumulated gas is released and germination follows. This could cause a spread of germination over a period of time, depending on the occurrence of damage to the seed coat.

However, Thornton (1944) showed that a large number of seeds, particularly of crop plants, germinate in carbon dioxide concentrations as high as 40–80 per cent, provided 20 per cent oxygen is also present. He concluded that for the seeds used, carbon dioxide does not inhibit germination in the presence of oxygen. In fact, Kidd also found that increasing oxygen concentrations and increasing temperatures decreased the inhibition caused by carbon dioxide, which is confirmed by Thornton. Under natural conditions a situation is rarely, if ever, met with, where carbon dioxide concentrations rise and yet the oxygen content remains high. Thus the conclusions reached by Kidd for *Brassica alba* seem to correspond more closely to events occurring naturally than the examples quoted by Thornton. In *Brassica* seeds the carbon dioxide was assumed to decrease the permeability of the testa, especially towards carbon dioxide. For *Cucurbita* (see Chapter 4) it has been shown that the seed membrane which controls gaseous diffusion is more permeable to carbon dioxide than to oxygen. Thus in *Cucurbita* it is possible that this permeability effect ensures spread of germination over a period of time.

In a number of seeds, small amounts of carbon dioxide can promote germination. The dormancy-breaking action on *Medicago* and *Trifolium* was mentioned in Chapter 4. In these plants this property may have resulted from a process of selection by agricultural practice rather than from natural selection. It seems possible that the sensitivity to small amounts of carbon dioxide causes rapid germination and that plants of this type are selected for, as this is often a desirable property in agricultural crops.

A suggested difference between weed seeds and the seeds of crop plants is their ability to survive while buried in soils for long periods. Under these conditions the seeds are liable to be exposed to high carbon dioxide concentrations while imbibed. Weed seeds are not damaged by these conditions and germinate rapidly when removed from the soils, as for example after ploughing. The seeds of many crops seem to suffer under these conditions. However, some weed seeds lose their viability quite rapidly if buried in the soil. For example *Scandix pecten*, *Linaria minor*, *Bartsia odontites* and *Polygonum aviculare* showed a reduction of 90 per cent in their viability in a two-year period (Brenchley and Warington, 1933). In the seed burial experiments of Duvel (1905) it was found that *Avena fatua* and *Lactuca scariola* survived under conditions where *A. sativa* and *L. sativa* lost their viability. Similar differences were observed between seeds from wild and cultivated plants of *Helianthus annuus*.

Relatively little direct information is available about the effect of oxygen on germination. Thus, no direct proof exists to show that seed membranes are impermeable to oxygen. However, indirect evidence exists for the testa of the upper seed of *Xanthium*. In these a rise in the external oxygen tension raises germination. In these seeds, therefore, the testa if not impermeable to oxygen, is at any rate only partially permeable to it. However, as already discussed, under natural conditions oxygen concentrations do not rise above 20 per cent. The ecological significance of these findings is not entirely clear. The germination of *Xanthium* seeds seems to occur only when a certain minimal threshold value of the internal oxygen concentration is reached. This results in the oxidation of an inhibitor present in the seeds. This inhibitor apparently prevents germination. Nevertheless, although there is a good correlation between external oxygen concentration and germination, it is not absolutely certain that germination is the direct result of destruction of the inhibitor (Wareing and Foda 1957; Wareing, personal communication). Under natural conditions the necessary threshold value is attained only very slowly due to the low permeability of the testa to oxygen. If the seed coat is punctured or damaged in some other way, these processes will be greatly accelerated. If the seed coat is completely impermeable to oxygen, then only damage can enable germination (Crocker, 1906; Shull, 1911).

From the above it is clear that no definite conclusion can be drawn about the function of gases in regulating germination. It seems probable that the internal concentration of gases in the seed is the only determining factor and it is possible that the carbon dioxide/oxygen ratio is determining rather than the absolute concentrations of each.

3. *Light*

An obvious advantage arising from the regulation of germination by light would be, in seeds having different light requirements, in adapting them to their habitat. Thus, light may prevent germination of seeds requiring light when they are buried under soil or leaf litter, or promote it when they fall on the soil surface. Such behaviour may determine how well a seedling will subsequently be able to establish itself. As already discussed, seeds can be divided into groups, viz. (1) those which require light, (2) those which are inhibited by light (Table 7.3 and see also Table 3.8) and (3) those which are indifferent to light. Ecologically speaking, it would be expected that all seeds requiring light for their germination belong to species of plants which necessarily germinate on the soil surface. All those seeds which are inhibited by light should belong to species of plants whose seeds will only germinate if they are covered by a certain amount of soil. However, this would be a gross over-simplification. Frequently short illuminations stimulate germination while prolonged illumination inhibits it, e.g. in *Amaranthus blitoides* and *Atriplex dimorphostegia*. Furthermore, seeds of the same genus may be light-requiring or light-inhibited, e.g. *Primula obconica*, which is light-stimulated, and *Primula spectabilis*, which is light-inhibited. In more extreme cases different varieties of the same species may be light-requiring or light-indifferent, for example *Lactuca sativa*. The possible value of a requirement for short illumination may be as follows. Seeds which are entirely

TABLE 7.3 — CLASSIFICATION OF SOME SEEDS ACCORDING TO THEIR LIGHT
REQUIREMENT
A – Seeds whose germination is stimulated by light
B – Seeds whose germination is retarded by light

A	B
Arceuthobium oxicedri	*Bromus sp.*
Daucus carota	*Datura stramonium*
Elatine alsinastrum	*Lycopersicum esculentum*
Ficus aurea	*Liliaceae*, various species
Ficus elastica	*Nigella sp.*
Gloxinia hybrida	*Phacelia sp.*
Gramineae, various species	*Primula spectabilis*
Gesneraceae, various species	
Lactuca sativa	
Lobelia cardinalis	
Lobelia inflata	
Loranthus europaeus	
Lythrum ringens	
Mimulus ringens	
Nicotiana tabacum	
Nicotiana affinis	
Oenothera biennis	
Primula obconica	
Phoradendron flavescens	
Raymondia pirenaica	
Rumex crispus	
Verbascum thapsus	

uncovered will fail to germinate as continuous light inhibits
them. However, if the seeds are partially covered, they may
be exposed briefly to light at certain periods of the day.
Under these conditions the seeds will germinate. In many
seeds light sensitivity is confined to only one part, e.g. in
Phacelia sensitivity is localized at the chalazal and micropylar
ends, in lettuce only the micropylar end of the seeds is light-
sensitive. Such sensitivity could perhaps ensure germination of
the seed only in certain orientations in the soil. It is possible
that these requirements are in some way related to the micro-

climate which is associated with certain conditions of illumination, which will affect subsequent development of the seedling. This is borne out by the complex interactions of light with other environmental factors which have already been discussed. There is no evidence, however, to show that localization of sensitivity has any ecological importance.

Light and temperature frequently interact. This interaction may be such as to make seeds sensitive to light only at certain temperatures but not at others. Some of the possible combinations of light and temperature are shown in Table 7.4. These interactions are so complex that it is almost impossible at present to interpret their ecological significance, if any. It must be borne in mind that very little is known about the micro-climate prevailing in the immediate vicinity of the seeds. In addition, the apparent combination of a number of factors,

TABLE 7.4 — CLASSIFICATION OF LIGHT – TEMPERATURE INTERACTIONS OF SOME SEEDS

(Compiled from data quoted by Stiles, 1950 and Koller, 1955)

Temperature / Plant	Light			Dark		
	Low	High	Alternating	Low	High	Alternating
Veronica longifolia	+	—	—	—	+	+
Epilobium hirsutum	+	—	—	—	+	+
Poa pratensis	+	—	—	—	—	+
Rumex crispus	+	—	—	—	—	+
Apium graveolens	+	—	—	—	—	+
Oryzopsis miliacea	+	—	—	—	—	+
Gesneraceae species	+	—	+	—	—	—
Ranunculus sceleratus	—	—	+	—	—	—
Chloris ciliata	×	+	—	—	—	—
Amaranthus retroflexus	×	+	—	—	—	—
A. lividus	×	—	—	—	—	—
A. caudatus	×	+	—	—	—	—

+ signifies germination is promoted by combination of treatments
— signifies no promotion
× germination inhibited by this set of conditions

revealed in the laboratory, may not in fact exist in nature. They may be no more than residual genetic properties which no longer have any direct survival value and which are retained as long as they have no harmful effect.

It seems possible that any ecological or survival value which light may have in regulating germination exists only in very special cases. Light requirement is frequently associated with small seeds, which are supposed to contain rather small amounts of reserve materials. It is frequently assumed that, because the seeds are small and therefore they contain little reserve materials, they must germinate under conditions where photosynthesis occurs very soon after germination. But seed size is no measure of the amount of storage materials present relative to the requirements of the seed. What determines the adequacy of the reserve materials is the ratio between the amount of these substances and the size of the embryo or seedling to be nourished by them. Moreover, a light requirement is by no means a universal phenomenon associated with all small seeds, although no large seeds are known to be light-requiring. The view outlined seems, therefore, to be an oversimplification of the ecological function of light.

A special case of adaptation exists in aquatic plants. In these, germination in deep water would be unlikely to lead to seedling survival. If the seeds are placed in a depth of water into which light can still penetrate, they will germinate and this will also occur if the water level falls, for example on river banks where the level varies seasonally or in swamps. This type of mechanism could explain the light requirement of seeds of *Typha latifolia*. Other plants in which this type of mechanism may be functioning are *Scirpus*, *Eichhornia crassipes* and *Juncus maritimus*.

Nevertheless, in general the light-requirement of many seeds is difficult to understand. Particularly as in many cases the amount of light required to stimulate germination is so small that it may be supposed that the seeds receive it under almost all normal circumstances. Even more difficult to explain is the function of the red, far-red mechanism in regulating germination. On information at present available it is impossible to see what possible ecological function is associated with this mechanism.

4. *Soil Conditions*

The nature of the soil also frequently affects germination. Some seeds respond to the calcium content of the soil in their germination behaviour, while others respond particularly to sodium content.

Seeds of *Hypericum perforatum* do not germinate in soils containing more than traces of calcium, whereas tomato seems to be indifferent to the calcium content of the soil.

Avena fatua germinates well in sandy loam and loess clay which also favour the growth of the seedling. In peaty soils germination is good but growth poor. Calcareous soils were found to be least suitable for germination among the soils examined (Bachthaler, 1957).

The high salt content of soils, especially of sodium chloride, can inhibit germination, primarily due to osmotic effects. In such saline environments the development of the seedling is extremely poor. However a number of plants, which have a special resistance to salt, can develop. This type of vegetation is frequently termed halophytic. Many of the plants so classified are distinguished by a high salt tolerance in various stages of their development, including germination, rather than by a positive salt requirement; high salt tolerance is not therefore proof that the plant is halophytic. But some plants show a definite requirement for a certain salinity and these germinate better in the presence of low concentrations of salt, and their subsequent development is also better. This is true for *Atriplex halimus* for example (Fig. 7.1). Even in this plant the tolerance to sodium chloride during growth and development of the seedling is ten to a hundred times greater than during germination. As a result, in many plants of this type, germination occurs when the salt content of the habitat has reached its lowest level, e.g. after rains. In a few plants occurring in a saline habitat vivipary is observed. The best known case is probably that of the mangroves, which grow for most of the year directly in sea water. It seems that during the process of germination these plants are much more sensitive to salt and are more easily damaged by it than the already grown seedling. In this case vivipary may be regarded as a means of evading the unfavourable environment during germination.

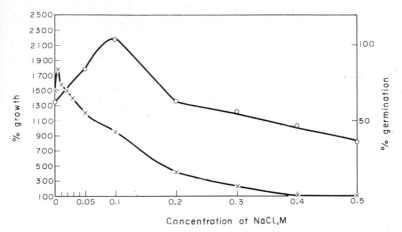

FIG. 7.1. Effect of NaCl concentration on growth and germination of
Atriplex halimus (Poljakoff-Mayber, unpublished)
× —— × % germination,
○ —— ○ growth, final dry weight as % of initial dry weight

Salt frequently accumulates in the organs of plants growing in saline habitats. As a result the fruits of the dispersal units of such plants often contain salt. In these plants germination is to some extent controlled by the salt content of the fruit. In these cases salt serves as an inhibitor of germination. In other seeds more specific inhibitors, not active through osmotic effects, are thought to be present.

5. *Inhibitors*

Many seeds do not germinate within their fruit, as discussed earlier. This observation, made very long ago, led to the idea that inhibiting substances are present in such fruits. In the course of time this idea was further extended, when it was found that extracts of such fruits also prevent germination. This finding supported the view that failure to germinate in the fruit was due to germination inhibitors and not due to unfavourable conditions, such as lack of oxygen or light, within the fruit. The prevention of germination of seeds in the fruit ensures that germination will only occur if the seeds are in some way dispersed. This will occur if the fruit decomposes, is broken or damaged or if it is eaten by animals and the seeds subsequently excreted.

In all these cases the dispersal of the seeds is increased and therefore the chances that the seeds will reach areas removed from the parent plant, where they are likely to survive, are improved. This view of the ecological function of fruits has recently been disputed (Gindel, 1960). Gindel claims that the presence of inhibitory substances has only been shown in fruits at a period of their development, when the seeds are not yet ready to germinate and the fruit is not completely ripe (in the biological rather than the culinary sense). According to this view, when the fruit is fully ripe, at the end of its development, many seeds will germinate in it. Examples are brought of cultivated plants such as apples, tomatoes and *Cucurbita maxima* and of a number of other plants such as *Ceratonia siliqua* (carob bean), *Laurus nobilis* and *Arbutus andrachne*. Accordingly it is supposed that during the development of the fruit the inhibitory substances present in it are converted to substances favouring germination, and that the fruit provides a nutrient medium and to some extent protection for the developing seedling. At present no evidence is available to indicate whether in fact inhibitors are decomposed during development of the fruit. Moreover, it is not clear what is the fate of the seedling after it has germinated within the fruit. Thus, no evidence exists whether the seedling remains in the fruit, without further development, until the latter decomposes or whether it pierces the fruit, as is the case, apparently, for carob beans. The case of mango may be relevant here. Among varieties of mango, viviparous and non-viviparous

kinds are known. Singh and Lai (1937) found that the viviparous varieties invariably showed splitting of the endocarp at
the broader end. This split endocarp was found to favour germination and thus might account for the phenomenon of vivi-
parity. In normal fruit with intact endocarp, radicle extrusion
may be prevented. How many seedlings survive this type of
germination is also not clear. At present this view must therefore be regarded as highly speculative and in many instances
in conflict with casual observation. The more usually accepted
view had led to a search for inhibitory substances in fruits.
Although these have in many cases not been identified,
nevertheless correlation between inhibitory power of extracts
and the germination behaviour of the seeds has been found. In
fleshy fruits inhibition may be correlated with sugar content.
In other instances salt content in the dispersal unit has been
found to be the cause of germination inhibition, as in *Zygophyllum dumosum*. In the case of *Zygophyllum* the dispersal unit
was found to contain salt in sufficient amounts to cause inhibition, but in addition weak inhibitory substances were also
present which accentuated this inhibitory action (Lerner *et al.*,
1959). In a few instances specific inhibitory substances have
been found to cause inhibition of germination. Among the inhibitors identified in fleshy fruits are parasorbic acid in *Sorbus
aucuparia* (Kuhn *et al.*, 1943), and ferulic acid in tomatoes
(Akkerman and Veldstra, 1947) hought this is probably not the
only inhibitor present, while in lemon, strawberry and apri-
cot, Varga (1957) showed the presence of a mixture of organic
acids which increased in amount as the fruit ripened. In dry
fruits the only inhibitor identified with any certainty is coumarin, in fruits of *Trigonella arabica* (Lerner *et al.*, 1959). Even
in *Trigonella* probably other, additional, weak inhibitory substances are also present. In the case of *Fraxinus excelsior* the presence of an inhibitor in the embryo itself has been shown by
Villiers and Wareing (1960). In most of these cases the function
of the inhibitors may be supposed to be that already mentioned, of dispersing the seeds in space and their germination
over a period of time. However, the mechanism may be even

more complicated. In seeds requiring chilling, the inhibitor may not cause a spread of germination in time. For example in *Fraxinus* and *Betula* the inhibitor probably only defers germination till a suitable time of the year, when chilling creates the necessary conditions which enable germination to proceed (Wareing, personal communication).

It must, however, be remembered that these presumed functions of inhibitors in fruits are by no means finally proven, and in fact they are very difficult to prove unequivocally. It is possible to interpret the observed facts differently (as already discussed). It is to be hoped that different approaches will lead to new lines of research into the probable biological and ecological function of inhibitors. More detailed examinations of extracts from fruits and seeds have already shown that these contain a mixture of substances, some of which inhibit while others stimulate germination, while yet others are active in affecting growth. The amount of these substances changes with time and with treatment of the seeds. It seems likely that germination is not simply controlled by inhibitors but that the interaction of both the promoting and inhibitory substances regulate it, as indicated by the case of *Fraxinus*, for instance (Villiers and Wareing, 1960).

6. *Biotic Factors*

Many other plant organs, other than fruits and seeds, contain inhibitors of germination. It has frequently been observed that leaves or leaf litter contain compounds which can inhibit germination of a number of seeds. The accumulation of leaf litter under trees could have a regulating effect on germination. If species differences exist in sensitivity to inhibitors, then those plants whose seeds are most sensitive to the inhibitor will not germinate in the immediate vicinity of plants whose leaves or straw contain such inhibitors. The presence of inhibitors in leaf litter might affect the distribution of certain plants, favouring some species and preventing the occurrence of others. It is often observed that some plants are entirely absent in the

immediate vicinity of certain other species. For example in
Israel the soil surrounding *Eucalyptus* trees is very poorly ve-
getated. Yardeni and Evenari (1952) showed that leaves of
Eucalyptus contained germination inhibitors. It was suggested
that the inhibitors present in the soil or leaf litter might be
responsible for the relative bareness of soil near *Eucalyptus*.
However, later experiments showed that although a very
strong inhibitor could be isolated from the leaves, the soil
from *Eucalyptus* groves did not inhibit germination of either
wheat or lettuce (Lerner and Evenari, 1961). Thus the inhibitor
in the leaves is either washed out or destroyed and no longer
exercises a biological function. However, it is possible that
very small amounts of inhibitor, together with the mechanical
effects caused by the leaf litter, may be responsible for prevention of germination. For example Dinoor (1959) showed that
the leaf litter under Valonia oak constituted a mechanical ob-
stacle to seedling establishment. When germination occurred
on the leaf litter, the roots failed to reach the soil and if it occurred underneath the litter, the shoot failed to emerge. Similar effects may occur in *Eucalyptus* groves or in other habitats
covered with leaf or straw litter.

Other causes for the absence of plants in the vicinity of
certain trees may be the competition for water and light, which
may have a more decisive effect than inhibitors. If competition
is very strong, then small amounts of inhibitors may also play
some part in regulating plant distribution.

Competition apparently is the crucial factor in establishment of *Juncus* species among the sward of grasses. The *Juncus*
can germinate and establish itself well when the soil is very
moist. However if the soil dries out a little, thus *Juncus* no longer germinates readily. Under these circumstances other plants
germinate and establish a cover which prevents light reaching
the *Juncus* seeds, which require light for their germination. The
Juncus is subsequently effectively prevented from germinating
(Lazenby, 1956). Such *Juncus* seeds retain their light sensitivity,
even if buried, for many years.

Inhibitors are present in the litter of a number of plants.

Thus in rice straw Köves and Varga (1958) showed the presence of a number of phenolic compounds which have inhibitory properties and suggest that these compounds may have some biological function. Beech litter also contains inhibitors. In this case inhibitors are absent in the litter immediately after the leaves are shed, but appear after it has been exposed to one winter (Winter and Bublitz, 1953). This could be related to the relatively brief viability of beech seeds, which germinate shortly after shedding and the seedlings establish themselves before the inhibitor develops in the litter.

Many members of the Cruciferae, such as *Brassica* and *Sinapis*, contain complexes of mustard oils in their fruit as well as in other parts of plants. In *Sinapis arvensis* the fruits contain mustard oils. Those seeds which are shed from the fruit germinate readily. The upper part of the fruit does not open readily and retains its single seed. This seed only germinates later, when the mustard oils have been washed out of the fruit. An interesting example of inhibitory action by other parts of the Brassica is provided by the case of *Brassica nigra*. Certain areas in California in the United States, which were damaged by fire, were resown with *Brassica nigra*. In the areas so treated many species failed to re-establish, while in other areas, not resown with *Brassica*, they germinated well and re-established themselves (Went *et al.*, 1952). Apparently, *Brassica* leaf litter contains an inhibitor which prevents the germination of certain plants. The presence of a water-soluble inhibitor in leaves of *Brassica nigra* was in fact proved (Scherzer, 1954).

In the aerial parts of *Echium plantagineum* two distinct inhibitors were demonstrated which are of special interest because they affected different species of plants quite differently. Differential effects were clearly demonstrated in this instance (Ballard and Grant Lipp, 1959).

The leaves of *Encelia farinosa* contain a powerful inhibitor of both germination and growth. A mulch of leaves of *Encelia* can inhibit the germination of a number of plants (Bonner, 1950). However, the suggestion that this accounts for the absence of plants very near to *Encelia* shrubs has been disputed by

Muller (1953). The latter claims that *Franseria* plants growing in a similar habitat also contain powerful inhibitors, yet its immediate environment is populated by other plants. Muller ascribes the difference between the ground cover in the immediate vicinity of these two shrubs as being due to differences in the growth habit and in the formation of leaf litter by them. Under *Encelia*, shrub-dependent herbs fail to develop because little leaf litter accumulates and the plant has a short life span. Under *Franseria dumosa*, however, leaf litter accumulates and therefore shrub-dependent herbs develop. Even under *Encelia*, open ground species develop readily.

A further instance of a plant containing a germination inhibitor is *Artemisia absinthium*, the inhibitor apparently being the alkaloid absinthin. A number of plants sown in the vicinity of *Artemisia* fail to germinate or to develop. For example, *Levisticum officinale* is killed up to a distance of one metre. Some plants are more resistant than others, *Senecio*, *Lathyrus clymena* and *Linum austrium* being extremely sensitive, while *Stellaria* and *Datura* are quite resistant (Funke, 1943). The ecological function is again in doubt, as seedlings even of sensitive plants, which survive for a year near *Artemisia absinthium*, subsequently develop quite normally, without further inhibition.

Many examples are found in botanical literature of the effects of plants on each other. Many of these are based on visual observation and most experiments have not been sufficiently rigorous to permit interpretation. Although it appears that plants do excrete substances or that substances leach out from them, which can affect other plants or seeds in their environment, the magnitude of the effects and their biological importance is still not clear. This whole problem has recently been reviewed by Evenari (1961).

The seeds of certain parasitic plants present a special ecological problem. Both *Striga* and *Orobanche* can develop only if the seeds germinate very near to plants which they can parasitize. The germination of these parasitic plants normally requires the presence of a stimulator, although they can be induced to germinate artificially. It appears likely that the stimulator which is required by these seeds is excreted or leached out from the roots of very many plants, and that it constitutes a normal metabolite in such plants. The requirement of *Striga* and *Orobanche* for such a factor in their germination would ensure that they germinate only under conditions where the chance to parasitize a suitable host is good. However, the apparent non-specificity of the stimulator, as compared to the specificity in the host requirement of these plants, throws some doubt on such an interpretation. Moreover, it appears that there is no one single stimulatory compound. For example, Sunderland (1960) was able to show in maize roots at least one water-soluble and a number of ether-soluble substances stimulating the germination of *Striga* and *Orobanche*. These substances act to some extent synergistically. It is possible that in different plants specificity effects are due to differences in the ratio of the amounts of such compounds acting synergistically. It remains to be seen whether this could cause certain species of parasites to germinate only near their specific hosts.

In the case of *Viscum album* and other *Viscum* species nothing is known about special germination requirements, the sticky nature of the fruit merely increasing the chances of it being carried from tree to tree by birds. This increases the likelihood of the seeds reaching a suitable host. The subject of germination of parasitic angiosperm seeds has recently been reviewed by Brown (1961).

Fire is another important factor which can control germination. This may be in one of a number of ways. Fire can remove vegetation and so improve light and aeration and at the same time remove competition for space, light and nutrients between the seedlings which are establishing themselves and the existing plants. Went *et al.* (1952) suggested that in addition fire can destroy accumulated inhibitors present in the soil cover and uppermost layers of the soil, again removing a possible cause of failure of seeds to germinate, or the seedlings to establish themselves.

Many animals can change the balance of different plants in a given area by grazing, by distributing the seeds, by the excretion of seeds after eating fruits, and by other means. Goats grazing in the Middle East denuded the area of forests and as a result caused the establishment of a different type of vegetation, containing many annuals and shrubs. Insects of various kinds which collect seeds may have certain local effects, for example ants accumulating certain kinds of seeds in or near their heaps.

Man can exercise and has exercised a profound effect on the distribution of plants in certain areas. Agricultural practice has always been designed to cause the establishment of certain plants to the entire exclusion of others. The disastrous results which this policy may have, was shown in the centre of the United States which was converted into an almost barren area for many years, the so-called "dustbowl". Jungle clearance in South America for temporary agriculture, without subsequent re-afforestation, is also resulting in man-made deserts. The methods formerly used to control vegetation were relatively simple, consisting of ploughing and weed eradication. However, modern usage of herbicides has greatly increased the means at the disposal of man to alter or control vegetation in all stages of development and consequently the danger of major changes over wide-spread areas has also increased.

BIBLIOGRAPHY

AKKERMAN, A. M. and VELDSTRA, H. (1947) *Rec. Trav. Chim. Pays-Bas* **66**, 441.

BACHTHALER, G. (1957) *Z. Acker and Pflanzenbau* **103**, 128.

BALLARD, L. A. T. and GRANT LIPP, A. E. 1959 *Aust. J. Biol. Sci.* **12**, 342

BONNER, J. (1950) *Bot. Rev.* **16**, 51.

BRENCHLEY, W. E. and WARINGTON, K. (1933) *J. Ecol.* **21**, 103.

BROWN, R. (1961) In *Handbuch der Pflanzen Physiologie*. **15** (In press) Springer-Verlag.

CROCKER, W., (1906) *Bot. Gaz.* **42**, 265.

DINOOR, A. (1959) M. Sc. (Agr.) Thesis, Rehovot (In Hebrew).

DUVEL, J. W. T. (1905) *U. S. D. A. Bureau of Plant Industry*, Bull. No. 83

EVENARI, M. (1961) In *Handbuch der Pflanzen Physiologie*. **16.**, 691 Springer-Verlag.

FUNKE, G. L. (1943) *Blumea*, **5**, 281.

GINDEL, I. (1960) *Nature, Lond.* **187**, 42.

JUHREN, M., HIESEN, W. H. and WENT, F. W. (1953) *Ecology* **34**, 288,

KIDD, F. (1914) *Proc. Roy. Soc. B.* **87**, 408.

KIDD, F. (1914) *Proc. Roy. Soc. B.* **87**, 609.

KOLLER, D. (1954) Ph. D. Thesis, Jerusalem (In Hebrew).

KOLLER, D. (1955) *Bull. Res. Council, Israel* **5**, D 85.

KOLLER, D. (1956) *Ecology* **37**, 430.

KOLLER, D and COHEN, D. (1959) *Bull. Res Council, Israel* **7**, D, 175.

KOLLER, D. and NEGBI, M. (1959) *Ecology* **40**, 20.

KÖVES, E. and VARGA, M. (1958) *Acta Biolog. Szeged*, **4**, 13.

KUHN, R., JERCHEL, D., MOEWUS, F. and MOELLER, E. F. (1943) *Naturwissenschaften* **31**, 468.

LAZENBY, A. (1956) *Herbage Abstracts* **26**, 71.

LERNER, H. R., MAYER, A. M. and EVENARI, M. (1959) *Physiol. Plant* **12**, 245.

LERNER, H. R. and EVENARI, M. (1961) *Physiol. Plant* **14**, 229.

MALL, L. P. (1954) *Proc. Nat. Acad. Sci. India*, **24**, Sec. B, 197.

MORINAGA, T. (1926) *Amer. J. Bot.* **13**, 126.

MORINAGA, T. (1926) *Amer. J. Bot.* **13**, 159.

MULLER, C. H. (1953) *Amer. J. Bot.* **40**, 53.

SCHERZER, R. (1954) M. Sc. Thesis, Jerusalem (In Hebrew).

SHULL, C. A. (1911) *Bot. Gaz.* **52**, 455.

SIFTON, H. B. (1959) *Canad. J. Bot.* **37**, 719.

SINGH, B. N. and LAI, B. N. (1937) *J. Ind. Bot. Soc.* **16**, 129.

SORIANO, A. (1953) *Rev. Invest. Agr.* **7**, 253.

SORIANO, A. (1953) *Rev. Invest. Agr.* **7**, 315.

STILES, W. (1950) *Introduction to the Principles of Plant Physiology*, Methuen, London.

SUNDERLAND, N. (1960) *J. Exp. Bot.* **11**, 236.

THORNTON, N. C. (1943–45) *Contr. Boyce Thompson Inst.* **13**, 355.

VARGA, M. (1957) *Acta Biolog. Szeged* **3**, 213.

VARGA, M. (1957) *Acta Biolog. Szeged.* **3**, 225.

VILLIERS, T. A. and WAREING, P. F. (1960) *Nature, Lond.* **185**, 112.

WAREING, P. F. and FODA, H. A. (1957) *Physiol. Plant* **10**, 266.

WENT, F. W., JUHREN, G. and JUHREN, M. C. (1952) *Ecology* **33**, 351.

WENT, F. W. (1953) *Desert Research Proceedings. Int. Symp.* 230–240 (Jerusalem Special publication No. 2, Res. Council, Israel).

WENT, F. W. (1957) *Experimental Control of Plant Growth* 248–251, Chronica Botanica Co., Waltham, Mass.

WINTER, A. G. and BUBLITZ, W. (1953) *Naturwissenschaften* **40**, 416.

YARDENI, D. and EVENARI, M. (1952) *Phyton* **2**, 11.

CHAPTER 8

REST PERIOD AND STATES OF DORMANCY IN OTHER ORGANISMS

THROUGHOUT the foregoing it has been stressed that the dry seed is characterized by a very low rate of metabolism. It has been pointed out that this reduced metabolic activity is closely associated with a reduction of the water content of the seed. However, as discussed in the chapter on dormancy, this low metabolism is not caused exclusively by lack of water, as the entry into and emergence from dormancy are often regulated by much more complex sets of conditions. The seed is not unique in showing dormancy and a reduced rate of metabolism over long periods of time, such phenomena being met with in other organisms too.

The partial reduction of metabolic activity occurs throughout the plant kingdom, as well as in bacteria, insects, crustaceans and even in mammals. The rest periods in all these groups of organisms seem to have one thing in common: a reduced rate of metabolism as compared with that during their active life. In the following we will attempt to point out those features which are common and those which are different in the rest period of the several groups, the relation between external conditions and the initiation and termination of the rest period in each group, and finally to point out some aspects of metabolism sometimes associated with rest.

Among the higher plants a rest period is found not only in seeds but also in buds, bulbs and various organs of vegetative propagation. In all these organs there is no dehydration as in seeds or at best only a partial dehydration, yet growth is discontinued until triggered off by some suitable external stimulus. Among the liverworts and mosses, growth and development can be arrested completely by drying. Yet when rehydrated, normal growth can be resumed. In the liverworts an alternative method of suspending growth and development, without in any way causing dehydration, is by photoperiodic conditions. Presumably following such treatment metabolism is also lowered or altered. In many algae a process of sporulation occurs, frequently under adverse conditions; in such spores, too, metabolism is greatly reduced. Among fungi two kinds of spores are met with, resting and non-resting. The latter germinate immediately under favourable conditions, while the former require some special conditions for breaking their rest. These spores are dormant. Resting spores are also formed in bacteria. In insect physiology the phenomenon of diapause is known. This again is associated with the reduction of the metabolic rates. In some cases a complete dehydration occurs in the larval stage. For example, in the larva of the midge *Polypedilum vanderplankii* the water content seems to be reduced to less than one per cent. In this dehydrated state the larva is very resistant to extreme conditions, such as dry heat at 102°C and low temperatures, such as those of liquid oxygen or nitrogen (Hinton, 1960). When such dehydrated larva are placed in water and a certain amount of water has been imbibed, they very rapidly begin to show metabolic activity. In contrast, the diapausing larvae of the sawfly (*Cephus*) absorb water whenever it is available, but their diapause is only terminated after chilling. In snails and crustaceans the phenomenon of a rest period or aestivation is not uncommon, but here the condition of rest is not usually associated with dehydration. It does sometimes occur, however, as in the eggs of the brine shrimp, *Artemia salina*, in which dormancy is associated with dehydration.

Among the higher orders of the animal kingdom, e.g. the mammals, a somewhat similar situation exists. Animals such as bats, certain mice, hedgehogs and even bears hibernate. This process again is characterized by a lowering of the rate of metabolism and in the case of the mammals by a lowering of the temperature of the animal.

From the above it follows that in some organisms rehydration alone is sufficient to end the rest period, while in others some additional treatment is required. Moreover, in all cases where the rest period exists without dehydration, some specific external factor is required for termination of rest, for example in *Lunularia*, which, as mentioned earlier, requires a suitable photoperiod for ending dormancy. Similar sequences are known for buds and bulbs. In insects, diapause is ended after exposure of the resting stage to certain changes in temperature, and may be induced by photoperiodic treatment (Lees, 1956 and 1959). In bacterial spores germination can be induced by a certain chemical treatment or by heat shock and possibly by mechanical treatment. In hibernating mammals, normal metabolism is resumed at certain periods of the year, although the precise triggering mechanism is uncertain. In some cases at any rate, photoperiodic response seems to exist.

Keilin (1959) has recently attempted to clarify some of the concepts and terminology associated with the rest period. According to him all states of lowered metabolism can be termed hypobiotic. Hypobiosis may result in hypometabolism (or dormancy), or in a lack of metabolism which Keilin calls cryptobiosis. The term of hypometabolism would encompass all dormant states, that is hibernation, aestivation, diapause and quiescence. Keilin's term cryptobiosis is synonymous with the earlier terms: latent life, anabiosis or abiosis. Keilin shows that hypobiosis can be caused by dehydration, lack of oxygen, cooling, high salt concentrations or combinations of these. However, to these we must now add a further condition, photoperiod.

Any attempt to understand the nature of the rest period must take into account two aspects, its induction and its termination.

The induction of dormancy is usually a result of changing external factors. Often it occurs when external factors are unfavourable to further development. However, unfavourable external conditions are not a necessary factor for induction of dormancy. The factors which induce dormancy are often also those which terminate it. Many of them may be due to changes which occur in the natural environment of the organisms. Internal changes are also sometimes involved in regulation of the rest period. However, the similarity in the behaviour of the various organisms is marked, particularly with regard to external conditions. These are common to all organisms living in a certain habitat and consequently are likely to serve as triggering mechanisms, determining the initiation or ending of metabolic processes in these organisms. The rest period may be induced or terminated as a result of such triggering mechanisms. External factors commonly involved are change in moisture conditions, change in temperature (diurnal or seasonal) and photoperiod. A number of examples will serve to show the general similarities in a wide variety of organisms.

The importance of water in inducing and terminating dormancy in many organisms has already been mentioned. Photoperiod, as already mentioned, plays an important part in the regulation of rest periods and is often linked to temperature. In buds of many woody plants a change of daylength from long to short days is accompanied by the beginning of the rest period, e.g. in *Betula*, *Fraxinus*, *Robinia* and others. In such plants, growth is greatly reduced in the whole plant and dormancy is induced in resting buds by short day treatment. If the plants are kept continuously in long day conditions, then dormancy may be absent and continuous growth maintained for long periods, (e.g. *Robinia*). In all these plants the beginning of dormancy occurs at a time of the year when the temperature falls, and the photoperiodic response is closely linked with the temperature (Wareing, 1956).

Dormancy of buds can sometimes be broken by a reversal of the dormancy-inducing photoperiod. Long photoperiods are especially effective in breaking the so-called summer dormancy. In *Pinus sylvestris* or *Quercus robur*, for instance, dormancy is readily broken by long day conditions. However, in many species in which winter dormancy has been induced (e.g. *Pinus*, *Quercus*), long day treatment is ineffective unless a chilling treatment has been given. Thus again photoperiod and temperature are closely linked (Wareing, 1956).

In the liverwort, *Lunularia cruciata*, growing in Israel, growth ceases entirely on exposure to long days, which normally occur in the dry spring and summer. Short days, which occur in winter (the period of growth of this plant), cause a resumption of growth (Nachmony, 1959).

In insects similar responses to photoperiod have been observed. In many arthropods exposure to long days prevents the onset of diapause, while short days induce it, for example, in *Acronycta* and *Metatetranychus*, in the Colorado beetle and in others. The exact daylength for preventing diapause is correlated with the latitude at which the organisms occur. The more southerly insects require a daylength of 13 hours while those from the extreme north require 20 hours of daylight for prevention of diapause. The reverse behaviour, prevention of diapause by short days is also known, for example, in the silk worm, *Bombyx mori* and the desert locust, *Schistocerca*. In insects, too, the photoperiodic response is closely linked to temperature effects. Generally speaking, high temperatures prevent the onset of diapause while low ones induce it. Again in those cases where long days induce diapause, high temperatures have the same effect, e.g. *Bombyx* (Lees, 1956 and 1960).

Emergence from diapause is not usually sensitive to photoperiod, but in *Dendrolimus* sensitivity to photoperiod is retained throughout diapause, and exposure of the dormant larva to long days induces a resumption of growth. Generally, in insects, the diapause is terminated by a change, usually a lowering, of temperature. The precise requirements for such changes are related to habitat; insects from cold climates require a lower temperature for ending diapause than those from warmer environments. The similarity of this requirement to the chilling requirements of seeds and buds is obvious.

As the rest period is always associated with a lower metabolism, induction and breaking of dormancy must be associated, directly or indirectly, with changes in the metabolism of the organism. It is probable that the changes which occur are extremely complex and are accompanied by the suppression of certain metabolic routes while others increase in significance. Possibly the enzymes present in the organism are also altered in some way. These changes may well be mediated by additional factors, unless it is considered that an environmental change can influence a variety of metabolic activities simultaneously. In many cases it has been shown that hormones do play a part in dormancy breaking and induction. In seeds and buds the presence of growth inhibitory and stimulatory substances has been repeatedly shown. In these it seems likely that the precise balance of such substances determines entry into or emergence from dormancy. Dormancy is often associated with the accumulation of inhibitors, but emergence from it is not necessarily accompanied by their disappearance; instead additional stimulators may appear.

In insects the situation is more complex. Growth, metamorphosis and also diapause are regulated by a system of hormones. These are secreted by neurosecretory cells, situated in various parts of the insect, such as the brain and the suboesophageal ganglion. These in turn affect the hormonal activity of the prothoracic gland and sometimes the corpora alata are also involved. Entry into and emergence from the diapause is determined by the activity of these hormonal systems. In diapausing insects it has always been found that some hormonal secretion is absent and that diapause can be ended by supplying the missing factor. On the other hand, no evidence has been found to show that the diapause is caused by inhibitory substances analogous to those occurring in plants tissues (Lees, 1961).

In fungal spores behaviour similar to seeds may be observed. For example the ascospores of *Neurospora* show a phenomenon of dormancy. These spores can be activated by heat shock (Goddard, 1935). This activation can be reversed by keeping the spores under anaerobic conditions for some time. Moreover chemical stimulation by furfural and its derivatives occurs in these spores (Emerson, 1948; Sussman, 1953; see also Cochrane, 1958).

A rather different situation exists in bacterial endospores. The formation of such spores is governed by factors quite different from those described so far for seeds, buds and insects. The process of sporulation is dependent on an adequate supply of nutrients. Bivalent ions, such as calcium and especially manganese are important, as is oxygen. Dipicolinic acid synthesis seems to be associated with sporulation. Temperature and pH during sporulation are essentially similar to those existing during normal growth. No evidence for photoperiodic control is available. The function of sporulation is by no means clear and it is apparently a normal event in many bacteria.

Spore germination seems to be somewhat difficult to define and various criteria have been used. Probably the most common one is that germination is characterized by a loss of heat resistance of the spores. Other changes also occur during germination, of which an increased oxygen uptake is reminiscent of events occurring in seeds. It seems to be an almost universal feature of germination of spores that they release dipicolinic acid (pyridine 2·6 dicarboxylic acid) which constitutes about 15 per cent of the spores. The release of dipicolinic acid, which is apparently closely associated with calcium metabolism, is accompanied by the loss of heat resistance of the spores. In addition, a peptide or peptides are released in appreciable amounts. Some workers differentiate between activation, which they regard as a reversible process, and germination which follows it. Other investigators suggest that spore germination is not an energy-requiring process and is primarily a kind of depolymerization reaction, possibly due to release of dipicolinic acid.

In common with many other organisms already mentioned, bacterial spores show a very rigid requirement for high moisture content for germination. A relative humidity of 95 per cent seems to be essential for germination to occur. Fungal spores appear to be less exacting in this respect. However, water is not sufficient to break the "dormancy" of such spores. A number of factors can cause germination. One of these, "heat shock", is similar to those met with in seeds or insects. Other factors causing germination are lalanine, certain ribosides, in some cases glucose and in others glucose with the addition of either alanine or a riboside.

Little is known regarding the mechanism controlling sporulation and germination. The dormant, heat-resistant spore contains only very few active enzymes, all of which are very heat resistant. The chief ones seem to be alanine racemase, catalase, adenosine deaminase and nucleotide ribosidase. However, additional enzymes are present in the spores which can be activated by treatments which stimulate spore germination, e.g. by heat shock. It is possible to show the presence of a very large number of enzymes in spore extracts, although these enzymes are not active in the intact spore.

It appears that during sporulation the activity of the glycolytic system and of the cytochrome electron transport system is greatly reduced, being totally inactive in the resting spore. In the germinating spores the utilization of glucose at first appears to be by direct oxidation. Later the glycolytic system becomes operative and the cytochromes become rapidly active. This is in some way similar to what has been suggested to be the situation in lettuce seeds (Chapter 6).

The problem of dormancy in bacterial spores has been discussed at an AIBS Symposium in 1957 (ed. Halvorson), and more recently by Halvorson (1961) and Halvorson et al. (1961).

The relation of the cytochromes to insect diapause is not clear. The cytochromes are considerably depressed, different ones being depressed to different extents. Nevertheless, cytochrome oxidase appears to be present in adequate amounts for the apparently very much reduced amount of other active cytochromes. It does not seem to be necessary at present to assume the presence of an alternative electron transport system during diapause.

A comparison of the rest periods in different organisms is exeedingly difficult. There is some indication that bacterial endospores constitute a distinct group in that neither their formation nor their germination is triggered by clearly defined external conditions. Only two factors are similar to those involved in insects and plants. This is the water requirement for germination and, according to some workers, the apparent water loss on sporulation, and the possibility of inducing germination by heat shock. The differences may, of course, only be apparent, because we do not know what the relevant triggering mechanism is. In contrast all the other groups mentioned, seeds, buds, fungal spores, insects and higher animals, clearly respond to certain more or less well-defined external factors. These are often clearly related to their growth and development. The rest period in these organisms can frequently be interpreted in relation to their life cycle, although such an interpretation may be open to doubt.

Despite the difference between bacterial spores and other organisms with respect to their response to external conditions there is some similarity at the metabolic level. This may be due to the similarities in basic metabolic processes in all living organisms. Therefore, a resemblance in the mechanism by which the same metabolic processes are lowered in various organisms may be expected. The question remains whether lowered metabolism results from dormancy or is its cause. We may be dealing with the inevitable result: metabolism is depressed because development is arrested and no demands are made on the sources of energy; alternatively, different signals or triggers depress metabolism and as a result development is arrested and dormancy results. These questions can only be answered when a great deal more is known about the fundamentals of dormancy and about many of the related phenomena.

BIBLIOGRAPHY

COCHRANE V. W. (1958) *Physiology of Fungi*, Wiley, New York.
EMERSON, M. R. (1948) *J. Bact.* **55** 327
GODDARD, D. R. (1935) *J. Gen. Phys.* **19**, 45.
GODDARD, D. R. (1939) *Cold Spring Harbor Symp. Biol.* **7**, 362.
HALVORSON, H. O. (1957) Editor, *Spores*, Publication No. 5, AIBS, Washington, D. C.
HALVORSON, H. O. (1961) In *Cryptobiotic Stages in Biological Systems*. Elsevier p. 32.

HALVORSON, H., O'CONNOR, R. and DOI, R. (1961) In *Cryptobiotic Stages in Biological Systems*, Elsevier. p. 41

HINTON, H. E. (1960) *Nature, Lond.* **188**, 336.

KEILIN, D. (1959) *Proc. Roy. Soc.* B **150**, 149.

LEES, A. D. (1956) *An. Rev. Entomology* **1**, 1.

LEES, A. D. (1959) Photoperiodism in insects and mites. In *Photoperiodism and related phenomena in plants and animals*, AAAS, Washington, D. C.

LEES, A. D. (1961) In *Cryptobiotic Stages in Biological Systems*, Elsevier, pp. 120 and 132.

NACHMONY, S. (1959) Ph. D. Thesis, London University.

SUSSMAN, A. S. (1953) *Amer. J. Bot.* **40**, 401.

WAREING, P. F. (1956) *An. Rev. Plant Phys.* **7**, 191.

INDEX OF PLANTS

AUTHOR INDEX

SUBJECT INDEX